Tentative Manual for Expeditionary Advanced Base Operations

Contents

Publishing Information	ii
Bibliographic Key Phrases	ii
Publisher's Note:	iii
Truth in Publishing: Disclosures	iv
Analytic Table of Contents: TENTATIVE MANUAL FOR EXPEDITIONARY ADVANCED BASE OPERATIONS	v
Abstracts	xv
TLDR (three words):	xv
ELI5:	xv
Scientific-Style Abstract	xv
Learning Aids	xvi
Mnemonic (acronym)	xvi
Mnemonic (speakable)	xvi
Mnemonic (singable)	xvi
References	xvii
Index-Idea-Generator: Innovative Indexes for Expeditionary Advanced Base Operations (TM EABO)	xvii

Publishing Information

Nimble Books LLC: The AI Lab for Book-Lovers

Fred Zimmerman, Editor

Humans and AI making books richer, more diverse, and more surprising

- (c) 2024 Nimble Books LLC
- ISBN: 978-1-60888-328-8

Nimble Books LLC ~ NimbleBooks.com

Bibliographic Key Phrases

Expeditionary Advanced Base Operations; EABO; Naval Expeditionary Forces; Marine Corps; Navy; Littoral Operations; Sea Denial; Sea Control; Competition Continuum; Force Design 2030; Joint Warfighting Concept; Stand-In Forces; SIF; 2030 Marine Littoral Regiment; MLR; Composite Warfare; Joint Logistics Enterprise; JLEnt;

Publisher's Note:

The world is in a period of intense geopolitical competition, with the potential for conflict looming large. As potential adversaries field ever more powerful weapons systems designed to keep US forces at bay, the US military must find new ways to project power and defend our interests. This manual, the *Tentative Manual for Expeditionary Advanced Base Operations* (TM EABO), outlines a bold new concept for employing naval expeditionary forces, operating in a highly distributed, mobile, and low-signature manner to "stand in" and deny adversaries control of key maritime terrain. This second edition builds upon previous work, drawing on lessons learned from war games, exercises, and experiments to provide a comprehensive description of how naval forces will conduct EABO across the competition continuum.

The document is a must-read for anyone interested in the future of naval warfare. It explores important topics such as:

- The operational context of EABO
- The planning and organizational considerations involved in conducting EABO
- The critical role of intelligence in supporting EABO
- The importance of information activities in creating and exploiting information advantages
- The unique challenges and considerations of aviation operations in support of EABO
- The essential role of logistics in sustaining EABO forces
- The concept of operations for conducting littoral operations

This manual is not just a theoretical exercise, but a practical guide for those seeking to understand and implement the EABO concept. It provides valuable insights into the future of naval operations and offers a roadmap for how the US military can effectively adapt to the changing security environment.

This annotated edition illustrates the capabilities of the AI Lab for Book-Lovers to add context and ease-of-use to manuscripts. It includes publishing information; abstracts; viewpoints; learning aids; and references.

Truth in Publishing: Disclosures

Context: "Tentative Manual for Expeditionary Advanced Base Operations" (TM EABO), 2nd Edition, May 2023, Department of the Navy, Headquarters, United States Marine Corps.

Strengths:

- **Ambition:** TM EABO is a bold attempt to codify a new way of fighting in the 21st century. It acknowledges that the "iron mountain" of traditional logistics is no more, and embraces distributed operations, information warfare, and a "stand-in" force posture.
- **Future-Focused:** The manual is full of buzzwords like "artificial intelligence," "machine learning," and "multi-domain operations," showing a keen awareness of emerging technologies and their impact on warfare.
- **Humor:** While it takes itself seriously, there are some amusing parts, like the dedication to the "enthusiastic proponent of expeditionary advanced base operations" who "challenged us to embrace disruptive thinking."

Weaknesses:

- **Bureaucracy:** It's a government document, so it's dense, repetitive, and prone to overusing jargon and acronyms. There's a distinct lack of compelling narrative, and the "single battle concept" is more of a mantra than a tangible strategy.
- **The "Tentative" Part:** The manual itself acknowledges that these ideas are still being tested. In other words, it's a work in progress, and some concepts might be more aspirational than practical.
- **Clichés:** Expect a healthy dose of "Think Outside the Box," "Embrace Disruptive Thinking," and "The Next Fight Will Be Different" as you navigate the manual's pages.

In Conclusion: This manual is a fascinating glimpse into how the Marine Corps is thinking about the future of warfare. It's ambitious, but also a bit rough around the edges. If you can wade through the jargon and clichés, there are some interesting insights to be found. But be warned: you might end up feeling like you just walked through a military-themed word cloud.

Analytic Table of Contents: TENTATIVE MANUAL FOR EXPEDITIONARY ADVANCED BASE OPERATIONS

Foreword: * Overview: Briefly summarizes the purpose and development of the manual, highlighting its iterative nature and the classified Concept for Expeditionary Advanced Base Operations (EABO). * Purpose: Outlines the three primary functions of the manual: education, experimentation, and driving action for future force development. * Scope: Defines the general characteristics, planning considerations, and organizational options discussed in the manual. Clarifies that it serves as a reference guide and not a cover-to-cover read. * Next Steps: Details the manual's role in validating and refining EABO capabilities, leading to its integration into formal naval doctrine.

Chapter 1: Introduction to Expeditionary Advanced Base Operations: * General: Defines EABO within the context of contemporary challenges, linking it to the Navy's Concept for Distributed Maritime Operations (DMO) and the Joint Warfighting Concept (JWC). Introduces the Concept for Stand-In Forces (SIF) and its relationship to EABO. * Operational Context: Discusses the evolution of force design assumptions and the challenges posed by adversaries' technological advancements and increased lethality. * Foundations of Expeditionary Advanced Base Operations: Defines EABO as a form of expeditionary warfare, outlining its core characteristics (stand-in, mobile, persistent, low signature, integrated, and cost-effective). Explains its purpose in the competition continuum and its connection to sea denial, sea control, and fleet sustainment. * Characteristics of Expeditionary Advanced Base Operations: Expands upon the core characteristics of EABO, highlighting their importance in countering adversary advantages and enabling joint force operations. * Expeditionary Advanced Base Operations Across the Competition Continuum: Details the application of EABO across the spectrum of conflict (competition, crisis, conflict) and its role in deterring aggression, shaping the operating environment, and supporting joint campaigns. * Relationship to Instruments of National Power: Explains the importance of coordinating EABO with all instruments of national power (diplomatic, informational, military, and economic) to achieve holistic effects and enhance national security objectives.

Chapter 2: Approach to Planning and Organization: * General: Emphasizes the need for bold action, decentralized execution, and a campaigning mindset to effectively employ EABO. Highlights the importance of planning for naval integration and formalized transitions between organizational structures. * Planning Context for Expeditionary Advanced Base Operations: Discusses the complexities of the operating environment and the need for detailed integration and coordination across echelons prior to conflict. Stresses the importance of de-escalation activities and the need for long-term thinking and coordination with allies and partners. * Inherent and Prescribed Conditions of Expeditionary Advanced Base Operations: Identifies inherent conditions (task organization flexibility, dispersion, resource competition, communication reliability, and sustainment) and prescribed conditions (host nation constraints, flexible objectives, and emissions control) impacting EABO planning. * Planning Framework: Outlines the key elements of the military decision-making model as applied to EABO, including integration, risk assessment, and the tenets of top-down planning, single-battle concept, and integrated planning. * Naval Command and Organizational Considerations: * Command Arrangements: Discusses the importance of centralized guidance, collaborative planning, and decentralized control in naval command arrangements. Defines the key roles of Navy and Marine Corps tactical forces. * Task Organization of Fleet and Maritime Forces: Explains how fleet commanders task organize forces into formations designed to operate across all dimensions of the maritime domain. Introduces the hierarchical structure of naval task forces (TFs, TGs, TUs, TEs). * Naval Task-Organization Hierarchy: Provides a detailed explanation of each level of task organization and their respective roles and responsibilities. * Framework for Decentralized Execution: * Mission Command and Control: Explains the principles of mission command and control and their application to EABO planning and execution. Discusses the importance of subordinates exercising initiative and understanding commander's intent. * Composite Warfare: Details the composite warfare framework, including its key personnel (OTC, CWC, warfare commanders, functional group commanders, and coordinators). Explains how composite warfare facilitates decentralized execution, collaborative planning, and command by negation. * Main Planning Considerations: Presents a comprehensive checklist of planning considerations for littoral force commanders, including assessing requirements and task-organizing EABs, evaluating warfighting functions, identifying warfare commander requirements, evaluating EAB posture, designating critical capabilities, identifying gaps and shortfalls, and assigning subordinate missions. * Planning Responsibilities: Outlines the planning responsibilities of the OTC, CWC, and the proposed EXWC. Defines the role of the LFC and highlights possible implementations. * Organization of Battlespace: Explains the role of battlespace control measures (FSCMs, waterspace management, and prevention of mutual interference) in coordinating naval force operations. Introduces the LOA as a control measure for integrating resources across domains. * Littoral Operations Area (LOA): Defines the LOA as a multidomain control measure, encompassing both landward and seaward littoral terrain, and its possible use as a subordinate maneuver space or a permissive control measure. * Considerations for LOA Planning and Development: Provides a detailed explanation of the LOA, its use in composite warfare, and the concept of sectors and engagement areas. * Naval Battlespace Terminology Related to Afloat Formations: Defines and explains the battlespace constructs used in composite warfare: surveillance area (SA), classification, identification, and engagement area (CIEA), and vital area (VA). * Control Measures: Emphasizes the importance of coordinating maneuver, fires, and airspace within the LOA through control measures. Introduces the Littoral Transition Point (LTP) as a control measure for surface littoral maneuver. * Command and Control: Discusses the dynamic nature of command relationships and the need for staffs to be prepared to execute transitions between command structures. Explains the concept of supporting situations (SUPSITs) and their application to EABO.

Chapter 3: Intelligence Operations: * General: Emphasizes the need for integrated intelligence operations to support planning, execution, and assessment of EABO. Introduces the importance of battlespace awareness, MDA, and establishing a baseline for the maritime operating environment. Highlights the benefits of partnerships with joint, interagency, and multinational partners. * Purpose and Scope: Explains the purpose of intelligence operations in providing an understanding of friendly, enemy, and neutral activities impacting EABO efforts. Clarifies that this chapter focuses on changes to how intelligence operations are conducted, not a change in function. * Intelligence-Led Operations: Introduces the concept of "actions to produce intelligence," emphasizing the shift from pursuing actionable intelligence to generating adversary activity to gain insight into their capabilities and vulnerabilities. * Naval and Joint Force Integration: Highlights the importance of close coordination and synchronization of effort between intelligence operations and the JFMCC, JFACC, and combatant command J-2 resources. Emphasizes the role of SOF in providing unique capabilities for information gathering and shaping the operating environment. * Operational Environment: Explains the need to understand the contested nature of the maritime domain, including the physical, human, and informational aspects of both the OE and IE. Introduces the JIPOE process and its application to EABO. * Contested Environment: Discusses the challenges of operating in a contested environment and the importance of intelligence operations in establishing a baseline picture of the OE. Highlights the role of intelligence operations in identifying escalations and preparing for armed conflict. * The Information Environment: Explains the complex nature of the IE, its potential impacts on the OE, and the need to analyze the physical, human, and informational aspects of both. Highlights the importance of understanding adversary capabilities in the IE. * The Littoral Environment: Defines the littoral environment and its importance to EABO planning. Explains the need to account for adversary capabilities and limitations across both segments of the littoral (seaward and landward). * Network Analysis and Civil Considerations: Explains the importance of network analysis and civil considerations (PMESII/ASCOPE) in gaining a comprehensive understanding of the OE. * Integrated Naval Intelligence Process: Discusses the need to integrate Marine and Navy intelligence efforts across multiple lines of effort, including employing integrated systems, ensuring system interoperability, integrating cross-domain solutions, training and exercising personnel, ensuring interdisciplinary proficiency, and synchronizing boards, centers, cells, and working groups. * Activity-Based Intelligence (ABI): Explains the ABI methodology, its potential benefits for EABO, and its challenges in a contested environment. Highlights the need for experimentation and training in degraded communications environments. * Support to the Sensing Enterprise: Discusses the concept of sensing as an enterprise service and its role in enhancing decision advantage for naval and joint forces. Highlights the importance of Marine collection platforms in extending the sensing network. * Collection Planning: Details

the importance of maintaining MDA, establishing a baseline for adversary and neutral activities, and using JIPOE to identify key terrain, adversary disposition and capabilities, relevant actors, target audiences, and key decision makers.
* Support for Effective Signature Management (SIGMAN): Explains the critical role of SIGMAN in preserving the survivability of EABO forces. Highlights the need to analyze adversary collection and targeting assets, understand their information and decision-making processes, and conduct conduit and emulative analysis.

Chapter 4: Information Activities in Support of EABO: * General: Emphasizes the importance of gaining and maintaining information advantage in EABO, using maneuver, fires, and information to influence friendly forces, allies, and adversaries. * Purpose and Scope: Defines information activities and their role in creating and exploiting information advantages (systems overmatch, prevailing narrative, and force resiliency) in support of EABO. * Information Environment Basics: * Adversary Activities in the Information Environment: Discusses adversary tactics in the IE, including targeting friendly systems, creating disinformation, and undermining US presence. Highlights the importance of assuming constant observation by adversaries. * Military Information Advantage: Explains the two mutually reinforcing elements of military power (physical combat power and military information power) and the importance of information advantage in gaining control over an opponent. * Information Warfighting Function: Defines the information warfighting function and its importance in managing and applying information to support the planning and execution of operations. * Assure Command and Control and Critical Systems: Discusses the importance of ensuring assured access to trusted information in a contested environment, outlining the challenges posed by adversary actions. * Provide Information Environment Battlespace Awareness: Highlights the importance of understanding the IE vulnerabilities, threats, opportunities, and their impact on operations. Emphasizes the need for gathering and fusing information from multiple sources. * Attack and Exploit Networks, Systems, and Information: Explains the use of traditional and information-specific capabilities to target adversary networks, systems, and information. Highlights the importance of technical target system analysis and the need for coordination with intelligence. * Inform Domestic and International Audiences: Discusses the importance of informing various audiences (domestic, international, and HN) to build understanding and support for EABO objectives and to counter adversary disinformation. * Influence Foreign Target Audiences: Explains the use of influence operations to shape perceptions and drive behavior change. Highlights the importance of understanding cultural, social, and political factors influencing target audiences. * Deceive Adversary Decision Makers: Defines deception operations and their objectives in misleading adversary decision makers. Highlights the importance of integrating physical actions with specialized capabilities. * Control Information Capabilities, Resources, and Activities: Emphasizes the importance of synchronizing information maneuver forces with all operations and the need for multidomain coordination and communication. * Creating and Exploiting Information Advantages: * Electromagnetic Spectrum Operations (EMSO): Discusses the use of EMSO to exploit, attack, protect, and manage the EMOE. Highlights the challenges of operating in a contested and congested EMS. * Cyberspace Operations: Explains the three distinct cyberspace operations (DODIN operations, DCO, and OCO) and their potential applications in EABO. Emphasizes the importance of protecting friendly networks and conducting OCO with caution. * Space Operations: Details the importance of space-based capabilities in supporting EABO. Highlights the challenges of operating in a contested space domain and the importance of coordinating with cyberspace and EMS planners. * Inform Operations: Defines inform operations and their role in correcting misinformation, establishing facts, and countering adversary propaganda. Emphasizes the importance of coordinating with higher headquarters and engaging key publics. * Influence Operations: Explains the use of influence operations to shape perceptions and drive behavior change. Highlights the importance of understanding cultural, social, and political factors influencing target audiences. * Deception Operations: Defines deception operations and their objectives in misleading adversary decision makers. Highlights the importance of integrating physical actions with specialized capabilities. * Information Maneuver Forces: Identifies the key information maneuver forces, including the Information Lead Planner, Multi-Domain Effects Team, CME, COMMSTRAT teams, Civil Affairs teams, PSYOP teams, and EW teams. Explains their respective roles and responsibilities in supporting EABO. * Alignment and Integration of Information in EABO: * Higher Echelon Alignment and Coordination: Emphasizes the importance of aligning information activities with higher-level objectives and coordinating with the Marine Corps information community.

* Naval Integration: Discusses the need to coordinate information activities with fleet objectives, highlighting the relationship between Marine information maneuver forces and Navy IW.
* Special Operations Force Integration: Emphasizes the role of SOF in generating, preserving, denying, and projecting information in support of EABO objectives.
* Authorities: Defines authorities as the power to act and their importance in enabling information activities. Discusses the distribution of authorities across echelons and the need to identify and address gaps in authority requirements.

Chapter 5: Aviation Operations: * General: Discusses the role of Marine Corps aviation in delivering lethal, effective, and survivable capabilities in support of EABO. Highlights the importance of the ACE in operating from austere locations and integrating with joint forces. * Purpose and Scope: Outlines the chapter's focus on discussing the roles, functions, and tasks of Marine Corps aviation in support of EABO. * Role of Aviation in Expeditionary Advanced Base Operations: Explains the importance of aviation in leveraging distributed effects and the need for a new concept to describe distributed aviation operations in support of EABO. Outlines the characteristics of aviation in support of EABO: persistent distribution, minimal sustainment, and networked command and control. * Functions of Aviation in Support of Expeditionary Advanced Base Operations: Explains the six functions of Marine aviation, highlighting the

unique requirements of operating in a contested littoral environment. Presents a proposed modernization of the Marine aviation functions to better account for the complexities of EABO. * Littoral Force Aviation Combat Element Supporting Relationships: Defines the relationships between the ACE and the littoral force, including general support, direct support, close support, and mutual support.
* Littoral Force Aviation Combat Element Relationships with the Joint Force: Explains the role of the ACE in supporting joint force missions, including air and missile defense, strike warfare, and the possibility of the ACE commander serving as a JFACC. * Littoral Air Command and Control: Outlines the role of MACCs in planning and executing aviation operations, highlighting their capabilities and limitations in supporting EABO. Discusses the capabilities and limitations of the Marine TACC, Navy TACC Afloat, Marine TADC, DASC, TAOC, EW/C Center, and MAOC in supporting EABO. * Aviation Ground Support: Explains the importance of AGS in supporting ACE operations, highlighting the key roles and responsibilities of the MWSS. Defines the six activities of AGS: forward aviation combat engineering, airfield operations, BRAAT operations, ADR operations, FARP missions, and ACSR operations. * Aviation Planning: Outlines the general approach to aviation planning in support of EABO, highlighting the need to balance efficiency, effectiveness, and flexibility in allocating scarce resources.
* Littoral Force Aviation Combat Element in Support of Joint Operations: Discusses the need to renegotiate the traditional allocation of long-range interdiction, reconnaissance, and air defense sorties in the context of EABO.
* Littoral Force Aviation Combat Element Liaison with Joint/Combined Entities: Explains the need for an integrated Navy-Marine Corps liaison element to effectively coordinate aviation operations with the JAOC/CAOC.

Chapter 6: Logistics Operations: * General: Defines the role of logistics in providing the resources, positioning, and sustainment of combat power. Highlights the importance of resilient and agile logistics in supporting EABO.
* Logistics in the Competition Continuum: Discusses the impact of the competition continuum on logistics, emphasizing the need for a proactive approach to prevent lethal engagement and maintain a competitive advantage.
* Tactical-Level Logistics: Emphasizes the importance of minimizing the traditional logistical footprint ashore and developing resilient supply webs that offer diversified distribution, redundant sourcing, and resilient infrastructure.
* Supply: Discusses the importance of forward provisioning techniques, the role of the GPN and MPF, and the use of OCS in reducing the logistical footprint and increasing flexibility.
* Maintenance: Highlights the need for a paradigm shift in maintenance support, emphasizing the use of simple, plug-and-play maintainable systems and the importance of local procurement and additive manufacturing.
* Transportation: Explains the need for an integrated approach to transportation, utilizing all modes of transport (sea, land, air) and both manned and unmanned systems. * Engineer Operations: Defines the role of engineer operations in supporting EABO, emphasizing their importance in establishing, maintaining,

and repairing infrastructure.

* Explosive Ordnance Disposal (EOD): Highlights the critical role of EOD in locating, engaging, and defeating explosive hazards across all domains in a contested environment. * Health Services Support (HSS): Explains the unique challenges of HSS in supporting EABO, emphasizing the need for an expeditionary HSS system and an aggressive preventive medicine program. Outlines the six principles of HSS (conformity, proximity, flexibility, mobility, continuity, and coordination).

* Medical Intelligence: Defines medical intelligence and its role in informing the LFC and medical planning.

* Medical Logistics Planning Factors: Outlines key considerations for medical logistics planning, including coordinating casualty evacuation, identifying forward-positioned stocks, and leveraging HN support. * Medical Management Planning Factors: Discusses the challenges of casualty management in a dynamic environment and outlines key considerations for the provision of medical services, including preventive medicine, casualty sorting (triage), patient holding, and patient evacuation. * Services: Highlights the importance of services in supporting EABO, particularly civil affairs, mortuary affairs, law enforcement, real property management, and the CCF. * Mortuary Affairs: Explains the challenges of mortuary affairs in a contested environment and the importance of coordination with the joint mortuary affairs command.

* Aviation Logistics: Discusses the unique requirements of AvLog in supporting ACE operations and the importance of integrating with the NAE. Highlights the importance of developing more distributed and resilient AvLog capabilities in the future. * Safety: Emphasizes the importance of safety planning in EABO and the integration of risk management controls.

* Operational-Level Logistics: Defines the core elements of operational-level logistics and emphasizes the need for collaboration with JLEnt, HN, coalition, and NGO partners. * Force Closure: Explains the concept of force closure and its importance in rapidly deploying and assembling forces. * Arrival & Assembly: Discusses the need for minimal arrival and assembly support for forces conducting EABO. * Intra theater Lift: Explains the importance of intra-theater lift in supporting distribution, sustainment, and maneuver. Highlights the use of both military and commercial assets.

* Littoral Movement: Explains the importance of littoral movement in supporting rapid deployment, sustainment, and maneuver of naval forces within contested areas.

* Medium Landing Ship (LSM): Introduces the LSM as a key platform for littoral maneuver and its role in supporting the 2030 MLR. * Medium Landing Ship Employment: Outlines the LSM's key capabilities and its role in supporting the sustainment of forces ashore. * Pioneer Battalion: Defines the Pioneer Battalion's role in conducting assured mobility operations across the littoral operating area, providing engineering and EOD support to enable littoral force maneuver and persistence.

* Pioneer Battalion Employment: Discusses the employment of the Pioneer Battalion in support of the MLR and its importance in establishing and maintaining

mobility across the littoral. * Theater Distribution: Explains the importance of theater distribution in coordinating, synchronizing, and prioritizing the fulfillment of requirements.

* Sustainment: Defines sustainment as a function that enables the functioning of military capability across an entire theater of operations. Highlights the importance of leveraging afloat and ashore logistics bases and establishing a "honeycomb" of logistical support sources. * Reconstitution & Redeployment: Explains the process of reconstituting and redeploying forces, highlighting the importance of maintaining combat capabilities and coordinating with USTRANSCOM and JLEnt agencies. * Component and Operational-Level Logistics: Outlines the responsibilities of Marine Forces component commands in supporting the sustainment of Marine units operating within the AOR, including the role of the FMFLC. * Strategic-Level Logistics: Defines the eight functions of strategic-level logistics (mobilization, procurement, war reserves, facilities, material readiness, strategic airlift and sealift, deployment and support, and force regeneration) and their importance in supporting EABO. Highlights the importance of compressing the levels of logistics to enhance resilience and responsiveness.

Chapter 7: Littoral Operations: * General: Provides a tactical and operational construct for planning EABO in support of littoral operations. * Concept of Operations: Emphasizes the importance of a concept of operations that deliberately manages the OE across the competition continuum, taking into account operational requirements, shifts in the OE, access/basing/overflight permissions, and adversary posture. * Plan of Execution: Outlines the three parts of the plan of execution: * Scheme of Maneuver: Describes how arrayed forces will accomplish the commander's intent. * Littoral Maneuver Plan: Covers the seaward and landward maneuver of forces to and within the LOA, detailing the use of multimodal transportation methods. * Plan of Supporting Operations: Details the elements of supporting operations that shape and establish conditions for executing the SOM, including information operations, host nation coordination, and reconnaissance. * Common Phasing Considerations: Identifies common phasing considerations for littoral operations, including shaping and reconnaissance, position selection and improvement, occupying the EAB, force protection and EAB security, and SIGMAN. * Shaping and Reconnaissance: Outlines the importance of conducting IPB, assessing adversary posture, surveying potential EAB locations, establishing EMS baseline, conducting counter-C5ISRT, and identifying mobility routes and hide sites. * Position Selection and Improvement: Emphasizes the need to select primary, alternate, and supplemental positions for EABs, balancing system ranges, proximity to vital areas, and threat levels. * Occupying the Expeditionary Advanced Base: Explains the importance of preserving tactical integrity and achieving optimal dispersion during embarkation. Outlines the role of the LMS in supporting littoral maneuver and the principles of combat loading. * Force Protection and Expeditionary Advanced Base Security: Discusses the importance of force protection in mitigating adversary actions and environmental conditions. Highlights the need for SIGMAN, camouflage, concealment, deception operations, and local site security measures.

* Signature Management: Explains the importance of SIGMAN in disrupting adversary target acquisition, tracking, and terminal guidance. Defines indicators, highlighting their characteristics (signatures, profiles, associations, contrasts, and exposures).
* Defensive Fires Plan: Discusses the planning for air and missile defense fires, highlighting the importance of integrating with external networks, employing tactical data links, and planning missile defense with the wider naval and joint forces.
* Mission Concepts of Employment: Presents CONEMPs for various systems and elements of the littoral force, including fires in support of surface warfare, fires in support of air and missile defense, operations in support of ASW, support to IW, and forward arming and refueling points (FARPs). * Fires in Support of Surface Warfare: Outlines the requirements for establishing a maritime fires kill chain, including networking, sensing, classification and identification of maritime objects, decision making, preplanned responses, and site selection. * Fires in Support of Air and Missile Defense: Highlights the importance of integrating with an external network, employing the common operational picture, and using tactical networks and data systems. * Operations in Support of Antisubmarine Warfare: Discusses the role of EABs in supporting ASW operations and the potential for emplacing sea mines. * Support to Information Warfare: Refer to Sections 4.5 through 4.8 for planning considerations and employment. * Forward Arming and Refueling Points (FARPs): Discusses the importance of FARPs in increasing the operational reach of aviation forces, adding resilience to aviation logistics, and increasing sortie generation rates.
* Forward Arming and Refueling Point Planning: Explains the three primary FARP methods (ground transported, air delivered, surface transported) and the factors to consider in selecting the appropriate method. * Forward Arming and Refueling Point Designs: Outlines the two primary FARP designs (assault and attack) and their respective requirements. * Fleet Interoperability: Emphasizes the importance of EABO in complementing the seagoing elements of the fleet and highlights potential missions for CSGs, ESGs, SAGs, ARG/MEUs, and combat logistics forces in supporting EABO.

Appendix A: Future Force Design and Considerations: * General: Provides a brief overview of the future force design considerations for the MLR and the infantry battalions that make up these units. * 2030 Marine Littoral Regiment: Discusses the 2030 MLR's role in conducting sea-denial operations and its core mission-essential tasks.
* Littoral Combat Team (LCT): Details the proposed 2030 LCT's composition, capabilities, and its role in commanding and controlling distributed EABs. * Combat Logistics Battalion (CLB): Outlines the CLB's core tasks, including prepositioning, ground supply, ground equipment maintenance, littoral transportation, general engineering, health services, operational contracting, and EOD. * Littoral Anti-Air Battalion (LAAB): Discusses the LAAB's composition, including elements from the MWSS, Marine wing communications squadron, Marine air support squadron, Marine air control squadron, and ground-based air

defense. * 2030 Infantry Battalion: Explains the 2030 infantry battalion's role in supporting naval and joint operations, emphasizing the need for organically equipped, networked, and mobile infantry units.

Appendix B: Mission-Essential Tasks: * General: Provides an overview of the revised EABO Marine Corps Tasks (MCTs) for the MLR and MEU, highlighting their role in informing readiness reporting and future development. * Proposed Marine Littoral Regiment Mission-Essential Task List: Outlines the core and core plus METLs for the MLR, Regimental Headquarters, LCT, CLB, and LAAB. * Proposed Maritime Expeditionary Unit Mission-Essential Task List: Outlines the proposed METL for the MEU Command Element, MEU Aviation Combat Element, MEU Ground Combat Element, and MEU Logistic Combat Element.

Appendix C: Experiment Objectives: * General: Outlines the purpose of experiment objectives in assessing the ability of units to execute operations according to the concepts laid out in the manual. * Missions that may be assigned during EABO: Lists the potential missions that may be assigned during EABO. * Experiment Objectives for Assessing Tasks and Missions: Provides a detailed list of experiment objectives and sub-objectives for assessing the ability of units to execute tasks and missions, highlighting the need for further analysis and planning to inform experimentation.

Appendix D: Abbreviations: * Provides a comprehensive list of abbreviations used throughout the manual.

Appendix E: Glossary: * Provides a comprehensive glossary of key terms and definitions related to EABO, naval warfare, and joint operations.

Abstracts

TLDR (three words):

Naval base operations.

ELI5:

This book is a guide for soldiers who are going to be working on a base far away from home. It talks about how to plan and work on the base, how to keep the base safe, and how to get supplies and things they need.

Scientific-Style Abstract

This tentative manual outlines pre-doctrinal considerations for conducting expeditionary advanced base operations (EABO). It details planning considerations and force organization for littoral forces operating from austere, temporary locations in contested maritime areas. The manual emphasizes the importance of decentralized execution, mission command, and naval integration. It examines the roles of various warfighting functions, including intelligence, information, aviation, and logistics, in supporting EABO. It also explores the need for resilience and adaptability in logistics, communication, and force protection. The document provides concepts of employment for various EABO mission sets and highlights the need for continual experimentation and assessment in order to refine EABO doctrine and best practices.

Learning Aids

Mnemonic (acronym)

EABO: **E**ngage, **A**dapt, **B**attle, **O**utmaneuver

Mnemonic (speakable)

Expeditionary Advanced Base Operations are **E**xcellent, **A**dvanced **B**ases for **O**perating.

Mnemonic (singable)

(To the tune of "Row, Row, Row Your Boat")

EABO, EABO, Sea denial's our goal, With low signature, We'll stay forward and bold!

References

Index-Idea-Generator: Innovative Indexes for Expeditionary Advanced Base Operations (TM EABO)

Index 1: Operational Environment & Logistics

A. Environmental Considerations

- 1. **Airspace**
    - 1.1 Air Defense
    - 1.2 Air Traffic Control
    - 1.3 Surveillance & Reconnaissance

- 2. **Cyberspace**
    - 2.1 Network Security
    - 2.2 Information Warfare
    - 2.3 Cyber Defense

- 3. **Electromagnetic Spectrum**
    - 3.1 EMCON
    - 3.2 Electronic Warfare
    - 3.3 Communications Security

- 4. **Information Environment**
    - 4.1 Propaganda & Disinformation
    - 4.2 Public Affairs
    - 4.3 Influence Operations

- 5. **Local Population & Environment**
    - 5.1 Cultural Considerations
    - 5.2 Humanitarian Assistance
    - 5.3 Civil Affairs
    - 5.4 Environmental Impact

B. Logistics & Sustainment

- 1. **Supply & Procurement**
    - 1.1 Forward Provisioning
    - 1.2 Cache Management
    - 1.3 Operational Contract Support

- 2. **Maintenance & Repair**
    - 2.1 Organic Capabilities
    - 2.2 Civilian FSRs
    - 2.3 Additive Manufacturing

- 3. **Transportation**
  - 3.1 Littoral Movement (LSMs)
  - 3.2 Airlift & Sealift
  - 3.3 Unmanned Systems

- 4. **Engineering**
  - 4.1 Base Camp Construction
  - 4.2 LTP Development
  - 4.3 Combat Engineering

- 5. **Health Services**
  - 5.1 Casualty Evacuation
  - 5.2 Patient Holding
  - 5.3 Medical Logistics
  - 5.4 Preventive Medicine

Index 2: Command & Control & Task Organization

A. Command Structures

- 1. **Composite Warfare**
  - 1.1 CWC & Warfare Commanders
  - 1.2 Functional Groups & Coordinators

- 2. **MAGTF**
  - 2.1 LFC as Task Group Commander
  - 2.2 Integration with Joint & Coalition Forces

- 3. **Supporting Relationships**
  - 3.1 General Support
  - 3.2 Direct Support
  - 3.3 Mutual Support

- 4. **Command & Control**
  - 4.1 Mission Command
  - 4.2 C2 in Contested Environments
  - 4.3 Communications Security

B. Task Organization

- 1. **Marine Littoral Regiment (MLR)**
  - 1.1 Headquarters & Fires
  - 1.2 Littoral Combat Team (LCT)
  - 1.3 Combat Logistics Battalion (CLB)
  - 1.4 Littoral Anti-Air Battalion (LAAB)

- 2. **Littoral Combat Team**
  - 2.1 Infantry Battalions & Companies
  - 2.2 Special Purpose Attachments
- 3. **Combat Logistics Battalion**
  - 3.1 Ground Logistics & Support
  - 3.2 EOD & Explosive Hazard Mitigation
- 4. **Littoral Anti-Air Battalion**
  - 4.1 Air Defense & Counter-UAS Capabilities
  - 4.2 Integration with Joint & Coalition Air Assets
- 5. **Aviation Combat Element (ACE)**
  - 5.1 ACE Functions (OAS, AAW, Assault Support, EW, etc.)
  - 5.2 Air Command & Control (TACC, TADC, MAOC)
  - 5.3 Air Traffic Control (MATC CO)
  - 5.4 Aviation Ground Support (MWSS)

Index 3: EABO Tactics, Techniques, & Procedures (TTPs)

A. Intelligence & Information

- 1. **Intelligence Collection**
  - 1.1 Baseline Development
  - 1.2 Activity-Based Intelligence (ABI)
  - 1.3 Target Systems Analysis (TSA)
- 2. **Signature Management**
  - 2.1 Own-Force Signature Assessment (OFSA)
  - 2.2 Electromagnetic Signature Control (EMCON)
  - 2.3 Counter-Reconnaissance
- 3. **Operations in the Information Environment (OIE)**
  - 3.1 Propaganda Countermeasures
  - 3.2 Public Affairs Strategies
  - 3.3 Information Warfare Techniques

B. Maneuver & Fire Support

- 1. **Littoral Maneuver**
  - 1.1 LSM & NGLS Operations
  - 1.2 LTP Selection & Improvement
  - 1.3 EAB Occupation
- 2. **Force Protection & Security**
  - 2.1 EAB Security Measures

- 2.2 Camouflage & Concealment
- 2.3 Deception Operations

- **3. Fires & Effects**
    - 3.1 Fires Kill Chain Development
    - 3.2 Targeting & Engagement of Maritime Targets
    - 3.3 Air & Missile Defense (AMD)

C. Logistical Support & Sustainment

- **1. Forward Provisioning**
    - 1.1 Cache Network Establishment
    - 1.2 Operational Contract Support (OCS)

- **2. Littoral Logistics**
    - 2.1 Surface Connector Operations (LSMs, etc.)
    - 2.2 Pioneer Battalion Employment
    - 2.3 Integrated Logistics Support

- **3. Aviation Support**
    - 3.1 FARP Operations & Design
    - 3.2 Aviation Ground Support (MWSS)
    - 3.3 Joint & Coalition Air Asset Integration

This index breakdown highlights the key areas of focus for EABO, offering a comprehensive overview for research and practical application. Each top-level category can be further subdivided for a more detailed exploration of specific topics and concepts. Remember, these are just starting points, and the. specific index structure will vary based on the specific needs and interests of the user.

TENTATIVE MANUAL FOR EXPEDITIONARY ADVANCED BASE OPERATIONS 2ND EDITION

MAY 2023

DEPARTMENT OF THE NAVY
HEADQUARTERS, UNITED STATES MARINE CORPS

DISTRIBUTION STATEMENT A: Approved for public release; distribution unlimited.
HEADQUARTERS
UNITED STATES MARINE CORPS
WASHINGTON, D.C. 20350-3000
PRIMARY REVIEW AUTHORITY: DC CD&I

RECORD OF CHANGES		
NUMBER	DATE	ENTERED BY

PCN 501 007704 00

INTENTIONALLY BLANK

DEPARTMENT OF THE NAVY
Headquarters, United States Marine Corps
Washington, DC 20350-3000

9 May 2023

FOREWORD

OVERVIEW

The *Tentative Manual for Expeditionary Advanced Base Operations* (TM EABO) was developed as part of an iterative process to test, refine, and codify the classified *Concept for Expeditionary Advanced Base Operations* (EABO), signed in March 2019 by the Commandant of the Marine Corps and Chief of Naval Operations, as well as to inform force design and development. This second edition of TM EABO includes updated information and captures lessons from war games, exercises, experiments, and other analysis to describe how naval forces will conduct EABO across the competition continuum. The information contained herein is therefore authoritative but not definitive; it provides the official baseline of ideas to be further tested and codified in doctrinal publications.

PURPOSE

Marine Corps concepts propose new and innovative approaches for addressing current or future gaps, shortfalls, or challenges for which existing methods or capabilities are ineffective, insufficient, or nonexistent. The original EABO concept must be read to understand and properly apply the new approaches called out in the TM. This manual was written to serve three primary functions:

   1) Educate the Fleet Marine Force (FMF) on the missions of, and the forces that conduct, EABO
   2) Facilitate live force experimentation to test and refine force structure and capabilities
   3) Drive action for future force development and serve as a foundation to move from learning to execution, including the expansion into formal naval doctrine

SCOPE

This manual describes the general characteristics and terms of EABO and provides planning considerations and options for force and battlespace organization. As a result of the dedicated efforts to implement Force Design 2030 over the past few years, the FMF has made great strides towards conducting EABO as envisioned within the approved concept. As an example, at the time of this writing 3rd Marine Littoral Regiment (MLR) within III MEF is at initial operating capability (IOC). This manual lays out factors and objectives for continued experimentation and assessment of force structure and capabilities associated with the MLRs, other task organized MAGTFs, and the naval vessels envisioned to support and sustain them. Included are considerations for command arrangements, as well as a series of cross-functional topics for exploration.

The tentative manual is meant as a reference manual; it is not designed to be read from cover to cover. It is recommended that all readers complete chapters 1, 2, and 7 along with their given functional interest in chapters 3-6.

- **Chapter 1:** Introduces Expeditionary Advanced Base Operations.
- **Chapter 2:** Addresses EABO planning and organizational considerations and provides a primer on the Navy Composite Warfare Commander (CWC) construct as an example for new tactical command relationships.
- **Chapters 3-6:** Explores functional considerations in the areas of Intelligence, Information, Aviation, and Logistics.
- **Chapter 7:** Focuses on utilizing EABO to support integrated littoral operations.

Based on feedback since its original publication in February 2021, notable changes, updates and inclusions have been made to this version including: expanding on the impact of irregular adversary forces and local populations within host countries; stressing the criticality of effective partnering including in the information environment; and aviation ordnance and forward arming and refueling considerations, especially in regards to the ability to service a variety of Navy systems that Marines are not currently trained or equipped to support. Most notably, the aviation and logistics communities have **extensively rewritten** chapters 5 and 6. These additions have made this version more comprehensive than the original, and it is our expectation that Marines and Sailors will continue to test and implement EABO ideas and capabilities based off this latest revision.

NEXT STEPS

EABO directly links with the Commandant's Planning Guidance and Force Design 2030 and aligns to four of the six USN Force Design Imperatives in NAVPLAN 22. This edition of the TM will be used by FMF units to conduct further live, virtual, and constructive force experimentation that validate, refine, and develop warfighting capability and generate best practices for tactics, techniques, and procedures as the concept moves beyond development to implementation.

Validated best practices should be incorporated into the Service's doctrinal publications as Marine Corps tactical publications (MCTPs) or Marine Corps reference publications (MCRPs). Marine Corps task revisions should codify EABO for core and assigned mission essential tasks across the FMF as applicable. Tested and validated solutions should be submitted for follow-on capabilities planning and entered into annual POM plan development. In the future, elements of all MAGTFs should be trained for sea denial operations.

Reviewed and approved this date.

KARSTEN S. HECKL
Lieutenant General, US Marine Corps
Deputy Commandant for Combat Development and Integration

PCN 501 007704 00

DISTRIBUTION STATEMENT A: Approved for public release; distribution unlimited.

This manual is dedicated to the late

Colonel Arthur J. Corbett, USMC (Ret.)

who served the Marine Corps Warfighting Laboratory for ten years after leaving active duty. He was a visionary, mentor, enthusiastic proponent of expeditionary advanced base operations, and good shipmate to all hands. He challenged us to embrace disruptive thinking and changing paradigms.

INTENTIONALLY BLANK

TABLE OF CONTENTS

Foreword i
Dedication iii

Chapter 1 Introduction to Expeditionary Advanced Base Operations 1-1
    1.1   General 1-1
    1.2   Operational Context 1-2
    1.3   Foundations of Expeditionary Advanced Base Operations 1-2
    1.4   Characteristics of Expeditionary Advanced Base Operations 1-4
    1.5   Expeditionary Advanced Base Operations Across the Competition Continuum 1-4
    1.6   Relationship to Instruments of National Power 1-5

Chapter 2 Approach to Planning and Organization 2-1
    2.1   General 2-1
    2.2   Planning Context for Expeditionary Advanced Base Operations 2-1
    2.3   Inherent and Prescribed Conditions of Expeditionary Advanced Base Operations 2-2
    2.4   Planning Framework 2-3
    2.5   Naval Command and Organizational Considerations 2-4
    2.6   Framework for Decentralized Execution 2-5
    2.7   Command and Control 2-12

Chapter 3 Intelligence Operations 3-1
    3.1   General 3-1
    3.2   Purpose and Scope 3-2
    3.3   Intelligence-Led Operations 3-2
    3.4   Naval and Joint Force Integration 3-3
    3.5   Operational Environment 3-4
    3.6   Integrated Naval Intelligence Process 3-8

Chapter 4 Information Activities in Support of EABO 4-1
    4.1   General 4-1
    4.2   Purpose and Scope 4-1
    4.3   Information Environment Basics 4-1
    4.4   Information Warfighting Function 4-2
    4.5   Creating and Exploiting Information Advantages 4-8
    4.6   Information Maneuver Forces 4-11
    4.7   Alignment and Integration of Information in EABO 4-13
    4.8   Authorities 4-14

Chapter 5 Aviation Operations 5-1

5.1	General	5-1
5.2	Purpose and Scope	5-1
5.3	Role of Aviation in Expeditionary Advanced Base Operations	5-1
5.4	Air Direction, Air Control, and Airspace Management	5-3
5.5	Functions of Aviation in Support of Expeditionary Advanced Base Operations	5-3
5.6	Littoral Force Aviation Combat Element Supporting Relationships	5-5
5.7	Littoral Force Aviation Combat Element Relationships with the Joint Force	5-6
5.8	Littoral Air Command and Control	5-6
5.9	Aviation Ground Support	5-9
5.10	Aviation Planning	5-10

Chapter 6 Logistics Operations ... 6-1

6.1	General	6-1
6.2	Logistics in the Competition Continuum	6-1
6.3	Tactical-level logistics	6-1
6.4	Operational-level logistics	6-13
6.5	Strategic-level Logistics	6-19
6.6	Conclusion	6-21

Chapter 7 Littoral Operations ... 7-1

7.1	General	7-1
7.2	Concept of Operations	7-1
7.3	Plan of Execution	7-1
7.4	Common Phasing Considerations	7-3
7.5	Mission Concepts of Employment	7-8
7.6	Fleet Interoperability	7-14

Appendix A Future Force Design and Considerations ... A-1

Appendix B Mission-Essential Tasks ... B-1

Appendix C Experiment Objectives ... C-1

Appendix D Abbreviations ... D-1

Appendix E Glossary ... E-1

LIST OF ILLUSTRATIONS

Figure 2-1. Notional naval task organization..2-4
Figure 2-2. Notional composite warfare organization...2-6
Figure 2-3. Littoral operations areas in the context of composite warfare..............................2-10
Figure 2-4. Notional littoral operations area...2-11
Figure 2-5. Navy supporting situations..2-13
Figure 3-1. All domain environment..3-4
Figure 3-2. The littoral environment..3-6
Figure 5-1. Proposed functions of Marine aviation...5-4
Figure 6-1. Sustainment web example...6-1
Figure 6-2. "Concentric Circle" Sourcing Logic..6-2
Figure 6-3. Spectrum of Forward Provisioning..6-3
Figure 6-4. Supporting capabilities..6-6
Figure 6-5. Notional force closure—advanced naval base through intermediate staging base...........6-15
Figure 6-6. Notional maneuver into littoral operations area...6-15
Figure 6-7. Conceptual models for levels of logistics..6-19
Figure 7 1. Notional concept of employment for maritime fires..7-9
Figure 7-2. Notional surface warfare unit of action delivers fires..7-10
Figure 7 3. Notional AAW unit of action intercept..7-11
Figure A-1. Notional Organization of the 2030 MLR A-1
Figure A-2. Notional Organization of the LCT A-2
Figure A-3. Notional Organization of the MLR CLB.. A-3
Figure A-4. Notional Organization of the LAAB A-4
Figure A-5. Notional Organization of the 2030 infantry battalion A-5
Figure A-6. Notional Organization of the 2030 infantry company.. A-5

INTENTIONALLY BLANK

CHAPTER 1

INTRODUCTION TO EXPEDITIONARY ADVANCED BASE OPERATIONS

1.1 GENERAL

In 2019, the Commandant of the Marine Corps and the Chief of Naval Operations approved the *Concept for Expeditionary Advanced Base Operations* (EABO), a foundational naval concept to address challenges created by potential adversary advantages in geographic location, weapons system range, precision, and capacity. It also created opportunities by improving our own ability to maneuver and exploit control over key maritime terrain, fully integrating Fleet Marine Force (FMF) and Navy capabilities to enable sea denial and sea control, and support sustainment of the fleet. EABO was tightly coupled and published in a planned sequence with the Navy's Concept for Distributed Maritime Operations (DMO) and is aligned to the ideas presented in the Joint Warfighting Concept (JWC.) Together, EABO and DMO advocate for integrated yet distributable naval formations to support sea denial and sea control in the face of potential adversaries who pose increasing challenges to current naval forces.

> The FMF conducts a variety of missions, most prominently afloat forward presence, crisis response, and all forms of amphibious operations. Thus, the FMF as a whole is *capable of* EABO rather than designed *exclusively for* EABO.

Since releasing the *Tentative Manual for Expeditionary Advanced Base Operations* (TM EABO) in 2021, the Marine Corps conducted multiple war games, exercises, experimentation, and other analyses to further develop and validate the central and supporting ideas. Other concepts, such as *A Concept for Stand-In Forces* (SIF) and *A Functional Concept for Maritime Reconnaissance and Counter-reconnaissance* [1] were approved to describe how Marines will be positioned forward at expeditionary advanced bases (EABs), shoulder-to-shoulder with our allies and partners, leveraging all-domain tools as the eyes and ears of the fleet and joint force. The SIF concept complements this version of the TM by describing capabilities and methods essential for forces conducting EABO. Designed to persist forward alongside allies and partners within a contested area, SIF can operate from EABs to leverage all-domain tools as the eyes and ears of the fleet and joint force. Stand-in forces complement the low signature of EABs with an equally low signature force structure. SIF have the enduring tasks of conducting reconnaissance and counter-reconnaissance at every point on the competition continuum and conducting sea denial when required in support of the naval campaign.

This updated tentative manual sets forth pre-doctrinal considerations for forces conducting expeditionary advanced base operations. Its provisions are applicable in varying degrees to all related situations, task organizations, tactics, techniques, and procedures. The specific missions, available means, and other variables of the operational environment (OE) will necessitate adjustments to the provisions as discussed in subsequent chapters and as we begin implementing EABO.

---

[1] Headquarters, US Marine Corps, *A Concept for Stand-In Forces* (2019); Deputy Commandant, Combat Development and Integration, *A Functional Concept for Maritime Reconnaissance and Counter-reconnaissance* (2022)

1.2 OPERATIONAL CONTEXT

For a generation, contemporary US force design and capability development modeled under three core assumptions: presumptive or readily achieved sea control, air superiority, and assured communications. However, as stated in Marine Corps doctrinal publication (MCDP)-1, "...advantages gained by technological advancement are only temporary, for someone will always find a countermeasure, tactical or itself technological, which will lessen the impact of the technology."

Continual rapid technological advancement and increases in the lethality, range and accuracy of potential adversaries' fielded weapons systems challenge US conventional military superiority and require the US military to continually reevaluate how it supports global power projection. Global competitors are increasingly fielding stand-off engagement capabilities - long-range systems designed to keep US forces out of key operating areas and push them farther from overseas allies and partners while minimizing risk to their own forces. This challenge is significant and cannot be met by merely refining current methods and capabilities - doing so would only delay the inevitable as adversaries only need to invest in slightly longer-range systems to regain the stand-off engagement advantage. Simply put, defeating this adversary strategy is not possible through an endless cycle of long-range one-upmanship, it requires a different strategy to regain the initiative.

Potential adversaries can also bring to bear advanced technologies during competition below armed conflict through irregular forces or proxies. These irregular forces can employ non-technical means to counter our conventional superiority, using elements of the population to gain information on Marine locations, composition, disposition, and strength. They can influence friendly and neutral networks in areas occupied by littoral forces to create unrest.

1.3 FOUNDATIONS OF EXPEDITIONARY ADVANCED BASE OPERATIONS

EABO provide engagement opportunities throughout the competition continuum and are a visible and tangible reminder of our Nation's resolve for friends and foes alike. Forces conducting EABO combine various forms of operations to persist within the reach of adversary lethal and nonlethal effects, changing their risk calculations. It is critical that the composition, distribution, and disposition of these forces limit the adversary's ability to target them, engage them with fires and other effects, and otherwise influence their activities.

> **EABO are a form of expeditionary warfare that involve the employment of mobile, low-signature, persistent, and relatively easy to maintain and sustain naval expeditionary forces from a series of austere, temporary locations ashore or inshore within a contested or potentially contested maritime area in order to conduct sea denial, support sea control, or enable fleet sustainment.**

Naval forces execute EABO throughout the competition continuum to deter aggression, set conditions within the theater before armed conflict occurs, and swiftly posture to fight within the maritime environment during a joint campaign. Advantageous force posture can be leveraged to disproportionately draw or distract enemy forces, or create dilemmas, which enable fleet forces to mitigate risk in a contested environment or seize opportunities elsewhere. The mobile and distributed nature of EABO imposes difficult choices upon the competitor and provides a force able to adapt and regenerate more quickly. The operating environment is likely one where the littoral force conducting EABO will be at a disadvantage in numbers of personnel and weapons, and proximity to interior lines. Additionally, living among, or near, the local population increases vulnerability to irregular threats from malign actors and adversary proxy forces. To succeed in this environment, commanders must promote an alert mindset that keenly balances risk to mission and risk to force, while seeking decisive engagement when it enables the fleet as part of the larger campaign.

The true advantage of EABO lie in the ability to support the projection of naval power by integrating with and supporting the larger naval campaign. Conceptually, naval expeditionary forces operating from the landward portion of the littoral, combined with the fleet's ability to operate seaward and in the airspace, in cyberspace, and in the electromagnetic spectrum (EMS) - give naval commanders the ability to operate in all five dimensions of the littorals (including: seaward [both surface and subsurface], landward [both surface and subterranean], the airspace above, cyberspace, and the EMS) in the maritime domain.[2] Given these organic capabilities, along with access to space-based capabilities, naval forces have the ability to gain and regain advantage in all-domain operations.

The desired end state for EABO is to contribute to integrated deterrence through Marine Forces that are structured and ready to persist, partner, survive, and fight effectively across an expanded maneuver space as a ready, capable, and combat-credible forward force. These forces will be capable of supporting the joint force commander (JFC) by:
- Establishing persistent sea denial capabilities forward to deter and, if necessary, blunt aggression in the littorals;
- Contributing to sea control;
- Conducting security cooperation activities to shape the operating environment by building partnerships, deter hostilities, counter malign behavior, and set conditions to achieve national security objectives;
- Contributing to fleet battlespace awareness;
- Supporting and, if directed, integrating with other joint, allied, and partner forces; and
- Refueling, rearming, and replenishing ships and aircraft in austere forward areas.

EABO missions include:
- Support sea control operations
- Conduct sea denial operations within the littorals
- Contribute to maritime domain awareness
- Provide forward command, control, communications, computers, combat systems, intelligence, surveillance, reconnaissance, targeting (C5ISRT), and counter-C5ISRT capability
- Provide forward sustainment to support and enable the joint force, and partners and allies

EABO tasks include:
- Conduct surveillance and reconnaissance
- Generate, preserve, deny, and/or project information
- Conduct screen/guard/cover operations
- Deny or control key maritime terrain
- Conduct surface warfare operations
- Conduct air and missile defense (AMD)
- Conduct strike operations
- Conduct antisubmarine warfare (ASW)
- Conduct sustainment operations
- Conduct forward arming and refueling point (FARP) operations
- Conduct security cooperation
- Conduct Irregular Warfare (IW)

---

[2] Office of the Chief of Naval Operations and Headquarters, US Marine Corps, *Littoral Operations in a Contested Environment* (Washington, DC: US Department of the Navy, 2017).

1.4 CHARACTERISTICS OF EXPEDITIONARY ADVANCED BASE OPERATIONS

<u>Stand-in</u>. EABO provide stand-in engagement opportunities throughout the competition continuum. During campaigning, forces conducting EABO engage allies and partners and their neutral civilian networks, counter threat networks, preserve access, and shape the theater for future operations. EABO also enables the persistent posturing of littoral forces within a potential adversary's tactical weapons engagement zone (WEZ) before conflict. During armed conflict, the combination of stand-in and stand-off engagement capabilities places the adversary on the horns of a dilemma: while the adversary seeks to discover and engage friendly stand-off forces, he exposes himself to the sensing, nonlethal, and lethal capabilities of stand-in forces.

<u>Mobile</u>. Forces conducting EABO have the organic resources and platforms sufficient to transit within a theater and conduct tactical maneuver across the seaward and landward portion of the littoral to accomplish assigned missions. Existing naval, joint, and allied/partner bases and stations in the theater also play an important role to project, sustain, and recover forces conducting EABO.

<u>Persistent</u>. Forces conducting EABO persist forward by moving with a high degree of flexibility within areas of key maritime terrain, presenting a light posture, sustaining themselves in an austere setting, and protecting themselves from detection and targeting. EABO diminish the reliance on fixed bases and easily targetable infrastructure.

<u>Low Signature</u>. Forces conducting EABO carefully manage signatures at all times, especially while conducting localized movement and maneuver. This allows them to remain positioned to achieve the desired operational effects while complicating adversary efforts to find and target them. During campaigning, utilizing military assets for movement and maneuver can make Marine forces easier to find and target by adversary irregular means and forces. Where feasible, forces should leverage host nation (HN) government and commercial assets to perform select support functions and reduce their reliance on external sustainment.

<u>Integrated</u>. The assigned mission sets within EABO are conducted within a joint and coalition framework, part of not merely an interoperable, but an integrated naval force. Task-organized Marine and Navy units project naval power through EABO by fusing their landward and seaward roles. For the purpose of this tentative manual, integrated naval units executing assigned tasks within and from EABs are referred to as littoral forces. Littoral forces do not connote a specific unit or formation. However, once task-organized as true blue-green teams, littoral forces embody the characteristics of EABO and persist within contested areas as they apply all available means to accomplish their missions.

<u>Cost-effective</u>. A stand-in force executing EABO is strategically cost-effective by virtue of its ability to undermine a potential adversary's cost-imposition strategy. Potential adversaries are investing in large numbers of comparatively inexpensive systems of adequate lethality, extended range, and greater precision to hold at risk the US military's expensive, sophisticated, and relatively few multi-mission platforms. Forces executing EABO are small, numerous, dispersed, relatively simple to maintain, and difficult to target, thus inverting an adversary's cost-benefit calculation when deciding whether to engage and upsetting the cost-imposition strategy.

1.5 EXPEDITIONARY ADVANCED BASE OPERATIONS ACROSS THE COMPETITION CONTINUUM

EABO provide value across the competition continuum through opportunities to conduct persistent engagement with partners and deter adversaries, which is a fundamental aspect of international relations. Within the aspect of cooperation, we undertake activities within EABO as a cooperative effort with like-minded nations during pre-conflict campaigning as a means of gaining and maintaining access, developing/enhancing allies' and partners' capabilities and increasing their interoperability with the joint force, countering malign behavior, and deterring regional aggression. The most common applications of EABO in this context involve contributing to regional surveillance to inform and support diplomatic, informational, military, and economic counteraction to violations of international norms. Cooperative activities may also include increasing familiarity with potential operating areas, collaborating in development and fielding of common equipment and materiel solutions, improving infrastructure, conducting exercises that build relationships and enhance collective warfighting capabilities, promoting deterrence, and supporting law enforcement against actions that violate HN or international laws.

Forces conducting EABO are designed to accomplish military objectives inside the WEZ, while decreasing risk to major fleet units, in support of the overall fleet concept of operations. In the event of crisis, naval forces conduct EABO to augment, enhance, or assist partner nations in defending sovereignty, controlling key maritime terrain, and contesting adversary *fait accompli* gambits.[3] In the event of conflict, naval forces conduct EABO to deny enemy freedom of action, impose costs, and shape the OE in support of integrated sea control and maritime power-projection operations.

1.6 RELATIONSHIP TO INSTRUMENTS OF NATIONAL POWER

Strategic competition requires coordinating efforts among all instruments of national power: *diplomatic, informational, military,* and *economic* (DIME).[4] These instruments must be mutually supporting, leverage all available capabilities across government, and contribute to the creation of effects in all domains. The Naval Service remains the preeminent US military component for sustained power projection, and a littoral force conducting EABO is a key enabler to a naval campaign. In the current environment, every action may affect multiple instruments of national power across the spectrum of conflict. Operational planning must consider these impacts, and the coordination among agencies and nations must be consistent and continuous.

1.6.1 Diplomatic

Diplomatic efforts can facilitate future EABO through mechanisms such as basing and staging rights, status of forces agreements (SOFAs), increased information sharing or other supporting HN agreements. These cooperative actions can also facilitate campaigning by providing diplomatic efforts with a forward-positioned US force that can reassure allies and partners, project power, develop HN capabilities, and provide credible deterrence options that enable discussions and negotiations.

1.6.2 Informational

EABO provide an opportunity to generate and project information within the information environment (IE), providing a means to convey intent, build relationships, promote partnerships, and undermine adversary efforts. By leveraging the information warfighting function, the commander of littoral forces enables strategic messaging and creates or exploits opportunities that support tactical and operational objectives. Information in EABO will be discussed further in chapter 4.

1.6.3 Military

The inherent mobility and persistent presence of forces conducting EABO enable joint force access and the ability to posture in international waters adjacent to friends, partners, competitors, and other actors. The effects produced by littoral forces are relevant for competition across the spectrum of conflict. These

---

[3] Per *Webster's II New Riverside University Dictionary*, a fait accompli is "an accomplished and presumably irreversible deed or fact."
[4] Joint Chiefs of Staff, *Doctrine for the Armed Forces of the United States*, JP 1 (Washington, DC: US Department of Defense, 2017).

effects influence the calculus of both friends and adversaries, by improving the US strategic position with increased presence in a contested area or by reducing force size in a given area.

1.6.4 Economic

The Naval Service may execute EABO in conjunction with US employment of the economic instrument of power. This requires collaborative planning with interagency partners as well as private enterprises. When cooperating with partners, the use of local contractors for logistical support can improve the US position in a region and counter a competitor's move to sideline US forces. Economic incentives can facilitate long-term security cooperation and ensure the availability of dual-use facilities such as sufficient harbors, docks, and bases. If planned effectively, these investments in foreign-nation infrastructure will potentially enhance US influence and set conditions for future operations.

CHAPTER 2

APPROACH TO PLANNING AND ORGANIZATION

2.1 GENERAL

The role of the commander of littoral forces, at every level, is to develop and exploit opportunities through bold action. Given human nature, commanders may have a natural inclination to employ larger formations to ensure mission accomplishment while enhancing self-defense. However, the virtues of smaller, more numerous, and more dispersed formations offer inherent force protection against an adversary with pervasive sensors and long-range precision weapons. Capitalizing on these attributes, commanders are able to employ well-placed, disciplined units across the littorals to achieve effects without resorting to traditional concentration of forces, which poses a credible threat to potential adversaries.

The complexity and danger found in EABO require the mental rigor to plan effectively and set the conditions for success not only during conflict but also before conflict begins. A littoral force is only relevant if it maintains the ability to apply force at the time necessary to generate options and influence the greater campaign. Operating in a distributed environment with limited support and resources will require an agile force with a critical-thinking mindset, demonstrating the mental agility to rapidly shift perspectives and generate alternatives.

> **Thorough planning and execution are critical to reducing risk to mission and risk to force. Leaders and Planners must consider safety and integrate risk management controls in every phase of planning and execution to help ensure successful EABO operations.**

Commanders must develop a campaigning mindset to effectively employ EABO not only in armed conflict, but also in competition and crisis. The cultivation of a campaigning mindset in planning, characterized by long-term thinking and coordination with allies and partners, increases political-military options while also presenting potential adversaries with increasing dilemmas. Campaigning across the competition continuum also demands more effective and complete naval integration.

Planning for, and possibly forming, naval task formations out of standing MAGTF organizations is a complex task. To achieve decision advantage it is essential to formalize and rehearse these transitions before a conflict occurs. To deliver the full potential of EABO and achieve desired effects, the traditional roles of blue and green components of a naval force must evolve. Command arrangements and functions must no longer restrict Navy components to seaward contributions and Marine Corps components to landward contributions to naval operations and power projection. Complete integration of naval forces conducting EABO requires an appreciation of the importance of integrating landward and seaward activities under fleet cognizance to achieve effects in all domains.

2.2 PLANNING CONTEXT FOR EXPEDITIONARY ADVANCED BASE OPERATIONS

The environment in which littoral forces will conduct EABO is both complex and dynamic. The commander must plan to operate across multiple domains and within the adversary's WEZ. Planning procedures require detailed integration and coordination at all echelons prior to conflict. Understanding the planning implications and requirements gained through analysis of the strategic environment, direction, and guidance is critical to ensuring mission success. Coordinating littoral force actions in the complex time, space, and geography of the littorals requires intricate planning, especially given the changing naval and joint force environment.

All planning efforts must seek to shape the general conditions of competition over an extended period of campaigning to be responsive to escalation and successful in armed conflict. It is important that planning factors include methods to de-escalate situations to prevent or deter armed conflict. De-escalation activities could include using intermediate force capabilities; generating, preserving, denying or projecting information; or other nonlethal options to counter adversary objectives while avoiding escalation.

2.3 INHERENT AND PRESCRIBED CONDITIONS OF EXPEDITIONARY ADVANCED BASE OPERATIONS

Littoral forces conduct EABO under the following inherent conditions and requirements:
- Task organization must be flexible, as task-organization requirements will vary depending on whether the forces are embarked, landing, or conducting operations within EABs ashore.
- Littoral forces must disperse as widely as possible to enable force protection and complicate the adversary's targeting cycle, while maintaining the ability to mass effects and impose cost upon the enemy in time and material.
- Widely distributed operations may create competition for limited shipping, connectors, tactical airlift, and assault support assets across the task force. Yet forces conducting EABO must rely on these for movement to the area of operations (AO) and maneuver within EABs. When movement means are insufficient to support planned operations and additional means cannot be made available, commanders must reevaluate and modify the scheme of maneuver (SOM).
- Littoral forces must carefully manage their signatures across spectrums and domains to enhance survivability while conducting localized movement and maneuver. Where possible, HN support may enable forces to reduce their signatures.
- Forces conducting EABO must be able to reliably communicate with higher echelons for information and intelligence despite enemy efforts to degrade or deny use of the EMS or in a degraded environment.

Also inherent to EABO and the maritime domain is the imperative to sustain the force, often from the sea. How the force is sustained inside the WEZ will require foresight and non-traditional methods, including forward provisioning, which is described in chapter 6. In many cases, forces must rely on globally positioned materiel, afloat and ashore, emplaced during competition, in order to enable and sustain operations until access to the joint logistics enterprise is established. This requirement, combined with mobility limitations imposed by terrain and infrastructure, may guide the commander to more heavily weight access to intermodal transfer points in positioning forces. Intermodal transfers represent periods of vulnerability that require close coordination among forces. Reliance on sustainment afloat through intermodal transfers may be a limiting factor for littoral force operations ashore. In some cases, sustainment will need to be conducted by and through a HN's military or civilian resources.

Prescribed conditions also influence and demand attention during planning for EABO in several ways. These conditions are usually, but not exclusively, prescribed by the establishing authority in the initiating directive in the form of constraints and restraints. First, the HN often imposes restrictions that limit the commander's freedom of action. For example, specific conditions may limit the allocation, employment, and control of surface attack munitions. HN prescribed constraints and restraints will influence the planning and operations of the littoral force in both competition and armed conflict.

Second, during planning for operations across the competition continuum, higher headquarters may initially assign the littoral force multiple potential objectives, with the selection of a specific objective delayed until just prior to execution. This requirement provides flexibility at the operational level by exploiting the littoral force's inherent mobility and flexibility. However, it complicates littoral force task organization and ship-to-objective-area planning. Separate plans for separate objectives must be prepared with normal attention to detail and, to the extent permitted by the various schemes of maneuver, must provide for the employment of the same littoral forces in the same general configuration and order. The planning challenge is similar to that faced by Marine Expeditionary Units (MEUs), which must be configured and embarked to execute multiple missions. Littoral forces must be configured and embarked so that they are capable of executing a variety of missions to achieve multiple objectives across the competition continuum.

Finally, higher headquarters may set conditions for strict emissions control, which creates a unique challenge during distributed operations. Given the character of EABO requiring extensive use of communications and electronic equipment to coordinate, direct, and support execution of EABO, widely dispersed EABs are potentially vulnerable to exploitation by adversary signal intelligence and electromagnetic warfare (EW) efforts. Special attention is required to ensure effective signature management (SIGMAN) and signal security during each phase, stage, and step of an operation.

2.4 PLANNING FRAMEWORK

Planning for EABO is built on the framework provided by the established military decision-making model. It is guided by the tenets found in the Marine Corps Planning Process/Navy Planning Process: *top-down planning*, *single-battle concept*, and *integrated planning*.[5] As with all planning, the enduring requirement is both continual refinement and the iterative nature of the process, with emphasis on branch plans and sequels. As stated in *Marine Corps Planning Process*, Marine Corps warfighting publication (MCWP) 5-10, planning should not be simply a series of steps. The incorporation of feedback loops and informed analysis is critical to allow the commander to function and thrive in an austere and dynamic environment. The following focus areas are specifically highlighted due to the unique strains that EABO will place on them.

Integration. *Joint Maritime Operations*, Joint Publication (JP) 3-32, outlines the importance of integrating maritime planning and ensuring consistency with the Joint Planning Process.[6] When conducting EABO, the need to function in a high-paced, resource-constrained environment adds importance to taking every opportunity to promote an integrated use of resources and capabilities across domains and through multiple echelons of command. The littoral force, operating in a mobile and distributed manner, will often need to leverage joint, coalition, and HN capability. In addition to integrating littoral force efforts with those of the greater joint force, the commander must seek these same efficiencies and integrate the operations and capabilities of the littoral force across time, space, and purpose.

Risk. Risk describes a situation involving exposure to danger or damage. Risk to force and risk to mission are inherent in all military operations. The littoral force commander (LFC) must also understand risk in terms of the opportunities that reside within the operating environment. Through understanding of the battlespace and appropriate planning, the commander manages risk through sequencing, phasing, and integration. Boldness is a key attribute in littoral operations.

---

[5] Headquarters, US Marine Corps, *Marine Corps Planning Process*, MCWP 5-10 (Washington, DC: US Marine Corps, 2020).
[6] Joint Chiefs of Staff, *Joint Maritime Operations*, JP 3-32 (Washington, DC: US Department of Defense, 2018).

2.5 NAVAL COMMAND AND ORGANIZATIONAL CONSIDERATIONS

2.5.1 Command Arrangements

Command arrangements include decisions made about how forces are task organized, what tasks each formation is assigned, what AO they are responsible for, who commands the different formations, and the command relationships among commanders. Naval command arrangements are based upon centralized guidance, collaborative planning, and decentralized control and execution. "Unity of command facilitates unity of effort. Unity of effort, the product of successful unified action, assures coordination and cooperation among all forces toward a commonly recognized objective, although they are not necessarily part of the same command structure."[7] When possible, naval tactical organizations seek to achieve unity of effort through unity of command.

Navy tactical forces provide operational commanders numerous capabilities through multi-mission platforms. Marine tactical forces provide operational commanders capabilities that complement those of Navy tactical forces and extend the reach of the fleet into both landward and seaward portions of the littorals.

2.5.2 Task Organization of Fleet and Maritime Forces

Naval task forces are normally delegated the authority to plan and execute tactical missions on behalf of the joint force maritime component commander (JFMCC), and they represent the highest echelon of task-organized naval forces. The fleet commander normally task organizes assigned tactical forces—including forces assigned to the FMF under the fleet commander's operational or tactical control—into formations with the capabilities to operate throughout all dimensions of the maritime domain to accomplish a given mission or set of missions. These formations may remain at the fleet level or be scaled to provide the right mix of capability and capacity through various combinations of *task forces*, *task groups*, *task units*, or *task elements*.

2.5.3 Naval Task-Organization Hierarchy

Naval task organization typically involves a tailored hierarchy of task forces (TFs), task groups (TGs), task units (TUs), and task elements (TEs), as depicted in figure 2-1.

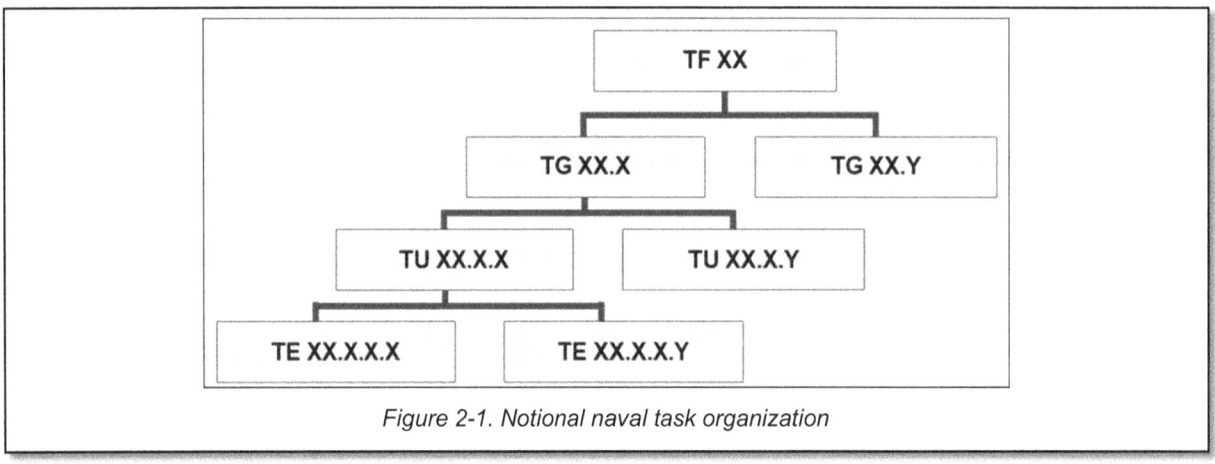

Figure 2-1. Notional naval task organization

---

[7] Office of the Chief of Naval Operations, *Maritime Operations at the Operational Level of War*, NWP 3-32 (Washington, DC: US Navy, 2008). For thorough discussions of unity of command and unity of effort in addition to *Maritime Operations at the Operational Level of War*, NWP 3-32, see Office of the Chief of Naval Operations, *Composite Warfare: Maritime Operations at the Tactical Level of War*, NWP 3-56 (Washington, DC: US Navy, 2015), and JCS, *Doctrine for the Armed Forces*, JP 1.

Three-Star/Three-Digit Task Forces. Navy numbered fleets and Marine expeditionary forces (MEFs) are the largest TFs in the maritime component, normally two-digit TFs. When employed in naval operations with multiple fleets or multinational partners, they are usually designated as three-digit TFs.

Navy Warfare Area and Functional Task Forces/Task Groups. Below each numbered fleet, subordinate commands are divided by naval warfare areas and functional areas. Although each fleet is slightly different in composition and organization, these are typically fleet battle forces, special warfare, maritime patrol and reconnaissance, logistics, undersea forces, naval expeditionary combat forces, and amphibious forces. These are identified as two-digit TFs in peacetime but may be re-designated as three-digit TGs for operational employment.

Fleet Marine Forces Organized as Naval Task Forces and Task Groups. When a MEF is assigned to the JFMCC to conduct EABO, major subordinate commands, such as the division, are normally designated as three-digit TGs.

Multi-mission Task Groups. These TGs are the largest mobile naval formations and may include carrier strike groups (CSGs), expeditionary strike groups (ESGs), surface action groups (SAGs), littoral combat groups, and littoral combat forces.

Task Units. Task units typically consist of smaller groups of ships and other naval assets that serve a functional purpose or are assigned a specific mission or limited range of missions in support of a multi-mission task group, in support of the fleet, or in support of a specific joint force function such as ballistic missile defense. Littoral forces will normally be task organized to conduct EABO as a TU and will normally be composed of forces at the O-6 level of command. These littoral forces may be subordinate to a multi-mission TG.

Task Elements. Task elements are often single ships. In situations involving a littoral force designated as a TU, its TEs would constitute the littoral force's smallest units of action.

2.6 FRAMEWORK FOR DECENTRALIZED EXECUTION

2.6.1 Mission Command and Control

The principles of maneuver warfare and mission command and control must permeate all actions of littoral forces conducting EABO, from planning through execution. During planning, commanders aim to create conditions during execution that enable subordinates to operate guided by the essential elements of mission command and control: *low-level initiative*, *commonly understood commander's intent*, *mutual trust*, and *implicit understanding and communications*.[8] Planning for EABO avoids a high degree of scripting and top-down direction, which usually aims to minimize uncertainties; rather, it must lead to understanding of the mission, intent, and broad guidance, creating freedom of action and maximizing opportunities for subordinates. Planning must be participatory, enabling leaders at every level within the littoral force to engage in the planning process and not merely consume a finalized and overly prescriptive directive. Given the anticipated OE, planning for EABO must ultimately foster a command and control (C2) environment which enables commanders at every level of the littoral force to cope with uncertainty, exercise initiative, generate tempo, and seize opportunities guided by mission and intent and bounded by a limited set of operational parameters.

Littoral forces may conduct EABO as part of either standing or temporary task forces. Given the anticipated OE, littoral forces are likely to find themselves dynamically re-tasked to support adjacent units and execute operations based on direction from outside the immediate chain of command. Thus,

---

[8] Headquarters, US Marine Corps, *Command and Control*, MCDP 6 (Washington, DC: US Marine Corps, 2018).

commanders must prepare their forces, at times, to respond to the control of external units that have been delegated authority to accomplish functionally aligned missions. Composite warfare is one example of a C2 construct that may enable operations within this fluid C2 environment. Operating within this construct requires coordination, planning, and procedures distinct from those typically familiar to Marine commanders and their staffs.

2.6.2 Composite Warfare

The commanders and staff of littoral forces must be thoroughly familiar with the contents of *Composite Warfare: Maritime Operations at the Tactical Level of War*, Navy Warfare Publication (NWP) 3-56. Composite warfare doctrine is a framework for command characterized by *command by negation*, *decentralized control and execution*, and *collaborative planning*. Due to the widely distributed nature of maritime combat, composite warfare employs command through preplanned actions to address threats by delegating warfare functions to subordinate commanders. Subordinates take action immediately, guided by the commander's intent, keeping the commander informed of the actions they take. Just as Marine commanders will communicate mission and tasks via operations orders updated by fragmentary order, composite warfare commanders issue orders via operational tasking message (OPTASK) updated by daily intentions message.

Key personnel within this construct include the officer in tactical command (OTC), composite warfare commander (CWC), warfare commanders, functional group commanders, and coordinators, as depicted below in figure 2-2. The OTC is the senior officer present eligible to assume command, and in application is often the fleet commander. The OTC may retain the duties of the CWC but will often assign these command functions to a subordinate. The OTC always retains responsibility for missions and forces assigned.[9]

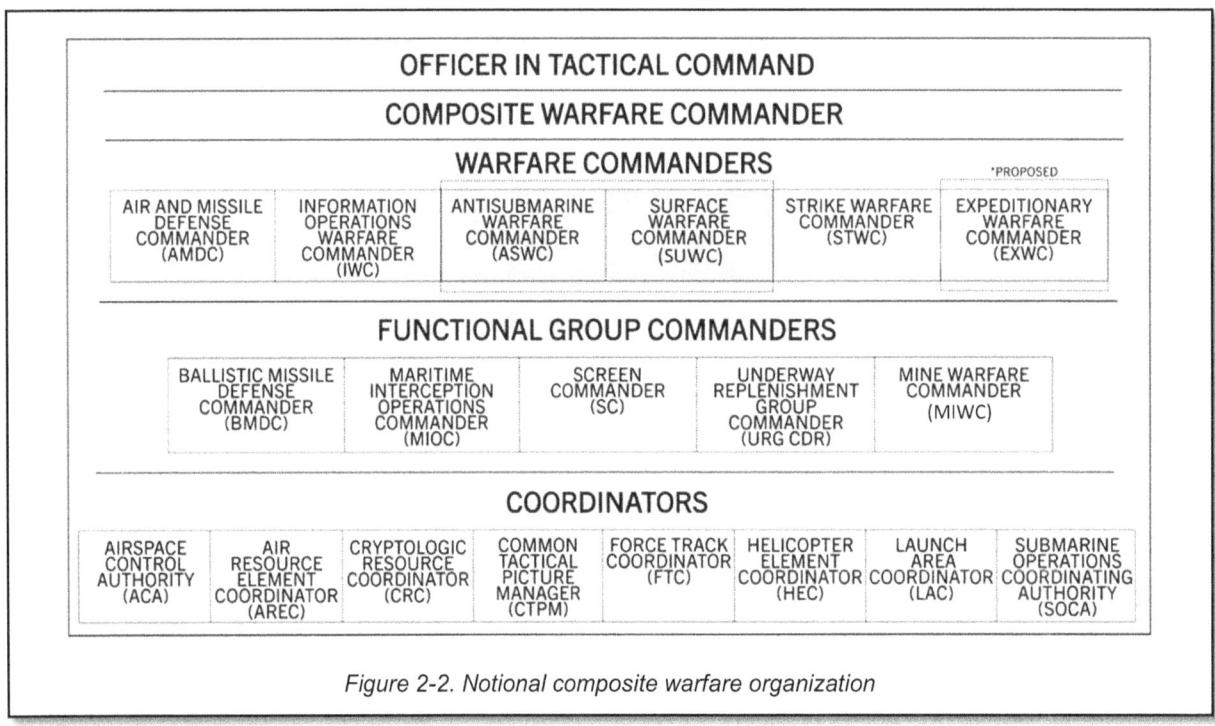

Figure 2-2. Notional composite warfare organization

---

[9] OPNAV, *Composite Warfare*, NWP 3-56.

The CWC delegates assignments as warfare commanders to subordinates. Warfare commanders are assigned duties of extended duration and broad situational applicability such as the air and missile defense commander (AMDC), surface warfare commander, and information warfare commander. The assignment of subordinate warfare commanders enables simultaneous offensive and defensive actions through decentralized execution. NWP 3-56 provides detailed descriptions of these positions and their responsibilities. The proposed expeditionary warfare commander (EXWC) is described in a draft Tactical Memo on the composite warfare construct to support further wargaming and experimentation.

Littoral forces should anticipate employment in a manner similar to a multi-mission ship or group of ships, extending the sensing range of the fleet and providing capabilities to warfare commanders in the surface, subsurface, and air domains.

The CWC may form temporary or permanent functional groups within the overall organization. Functional groups are subordinate to the CWC and are usually established to perform duties that are generally more limited in scope and duration than those conducted by warfare commanders. In addition, the duties of functional group commanders generally span assets normally assigned to more than one warfare commander. The ballistic missile defense commander is an example of a functional group commander that may be supported by littoral forces.

Finally, resource coordinators may be established to execute the policies of the CWC and respond to the specific tasking of either warfare commanders or functional group commanders. Resource coordinators are usually designated when specific resources impact more than one warfare commander. The air resource element coordinator is an example of a coordinator who executes the policies of the CWC with all providers of air resources.

2.6.3 Main Planning Considerations

The following planning considerations allow the LFC to design the littoral force appropriately to accomplish assigned tasks.

Assessing Requirements and Task-Organizing EABs. Forces organized for EABO may be assigned a single task, a few tasks, or many tasks depending on requirements of the mission. Once the requirements of an EAB are established, the commander task organizes elements of the littoral force to accomplish the mission. By designing a purpose-built task element consisting of a unit of action (typically a reinforced platoon), supporting staff, and sustainment, the commander best supports mission requirements while minimizing signature. These task elements do not exist as a permanent unit in force structure – they are formed as needed. The littoral force's units of action should be organized based upon the commander's analysis of mission requirements.

Warfighting Functions. The commander must conceptualize capabilities to execute operations in terms of the warfighting functions. Considering the assigned tasks through the lens of the warfighting functions encourages planning for the design and employment of forces in the most ideal posture to achieve desired effects across the competition continuum.

Warfare Commander Requirements. The commander of littoral forces may be designated an EXWC. As an EXWC, the LFC may be delegated authority and resources to accomplish missions assigned by the CWC. Simultaneously, the EXWC would retain the requirement to support hierarchically adjacent warfare commanders in support of their assigned missions within respective domains. The EXWC must also be aware and capable of executing relevant preplanned responses as prescribed in the OPTASK.

Evaluating EAB Posture. The fundamentals of offensive and defensive planning provide useful considerations for the commander to integrate into his/her planning model. These may include *flexibility*, including the desire to maintain multiple courses of actions (COAs); *mutual support*, where the relationship and positioning of units mitigate gaps that exist when units operate independent of each other; and *surprise*, where the commander employs available capabilities to deceive the adversary and manages the signature of his/her forces to present a desired posture.

Designating Critical Capabilities. Based on the mission and the commander's assessment of the threat, the commander must determine the critical capabilities of the littoral force. Some of these critical capabilities may result from the littoral force's role and functions within the composite warfare organization and the demands of the various warfare commanders. The commander organizes the force to fully capitalize on these capabilities and ensure responsiveness within composite warfare.

Identifying Gaps/Shortfalls. Throughout all phases of the planning process, gaps and shortfalls are reevaluated and assessed in terms of risk to the force and risk to the mission. Based on risk determination, gaps and shortfalls are communicated via the chain of command. This information may cause assigned tasks to be modified, allocation of additional resources or modification of the force posture.

Assigning Subordinate Missions. Having considered the requirements of the warfare commanders and joint force, capabilities of the adversary, impacts of the local environment, and requirements to sustain the force, the commander is prepared to organize the littoral force to conduct operations. The littoral force commander assigns subordinate forces and missions and, perhaps most importantly, communicates the capabilities of the task-organized force to the CWC and warfare commanders.

2.6.4 Planning Responsibilities

Commanders must understand the different levels of authority and the impact each has on the commander's ability to control assigned and attached forces. Commanders provide tactical direction and guidance through a clear statement of intent. The nature and focus of planning varies by echelon, while all actions are coordinated through the lens of single battle. To achieve unity of effort, commanders ensure that (1) subordinates clearly understand the command authority they have been granted and (2) the forces assigned understand what this authority allows.

Officer in Tactical Command. The OTC is the senior officer present eligible to assume command or the officer to which the senior officer present has delegated tactical command. The OTC's planning will normally focus on power projection and sea-control operations.

Composite Warfare Commander. Appointed by the OTC, the CWC's planning efforts will normally focus on operations to counter threats to the force. The CWC appoints warfare commanders who in turn align resources to surveillance areas (SAs); classification, identification, and engagement areas; and vital areas[10], which are discussed below in subsection 2.6.5.

Expeditionary Warfare Commander (proposed). The proposed EXWC is the senior commander of littoral forces who is subordinate to the CWC for the execution of assigned missions. As a warfare commander, the EXWC simultaneously possesses certain delegated authorities of the CWC, while also supporting hierarchically adjacent warfare commanders in the execution of their assigned missions. Thus, the EXWC's planning efforts must address his/her own operational requirements, while also planning to

---

[10] OPNAV, *Composite Warfare*, NWP 3-56.

support those of other warfare commanders. If the EXWC is assigned a littoral operations area (LOA),[11] this commander may also be assigned authorities of the littoral force commander, discussed in the next paragraphs and therefore be responsible for several primary decisions during planning, which are discussed below in subsection 2.6.5.

For the purposes of experimentation, this manual uses the general term LFC to describe the individual who, regardless of the task organization or echelon, exercises command over all littoral forces conducting EABO within an LOA. While not exhaustive, the following examples illustrate some possible implementations:
- A naval O-6 task-unit commander, subordinate to a task group, may fulfill this role as the senior commander of a littoral force.
- A naval O-8 task-group commander may fulfill this role, exercising command over littoral forces within multiple LOAs.

The rationale for introducing the term LFC is twofold. First, during EABO an LFC may not operate within the composite warfare construct and therefore not fulfill any composite warfare roles. Second, an LFC operating within composite warfare may *not* be designated a warfare commander, function group commander, or coordinator.

2.6.5 Organization of Battlespace

When employed in naval operations with multiple fleets or multinational partners, numbered fleet and MEF commanders are designated as three-digit task forces by the JFMCC. A likely construct for naval-force employment is that these three-digit task forces will be assigned AOs by the JFMCC, and each will serve as OTC within the assigned AO. Task force commanders will organize and manage their battlespace according to doctrinal maneuver control measures, fire support coordination measures (FSCMs), waterspace management, and prevention of mutual interference.

When conducting EABO, task force commanders must take advantage of littoral terrain to integrate with joint force operations and generate tempo in decision making and action against the adversary. Maneuver in the littorals creates the possibility to extend the range of fleet sensors and shooters beyond the classification, identification and engagement areas and SAs of traditional task groups. Accordingly, this manual allows for task force commanders, and the task groups supporting them, to experiment with appropriate naval command and control to best enable integrated, all-domain operations for modern naval warfare. Task force commanders can employ littoral forces to conduct EABO at any echelon (TE, TU, TG, and TF), as required by mission and geography. Task group commanders may be designated as CWCs, and littoral forces designated as task units and below may operate under task group command using composite warfare. Marine Corps units in those formations must be able to integrate seamlessly with the CWC structure. Task force commanders may also integrate littoral forces conducting EABO with adjacent task groups using other battlespace management constructs.

For purposes of simplicity, these various options can be compressed into three general types of command arrangements for experimentation, summarized as follows:
- Littoral forces operate under the CWC of an afloat Navy task group
- Littoral forces operate as their own task group using Marine air-ground task force (MAGTF) C2
- Littoral forces operate as their own task group using composite warfare

---

[11] Discussed in greater detail in subsection 2.6.5, an LOA is a geographical area of sufficient size for conducting necessary sea, air, and land operations in order to accomplish assigned mission(s) therein.

Figure 2-3 provides examples of a maritime AO integrating littoral forces with other forces of the fleet operating in the air, sea, undersea, and in space and cyberspace.

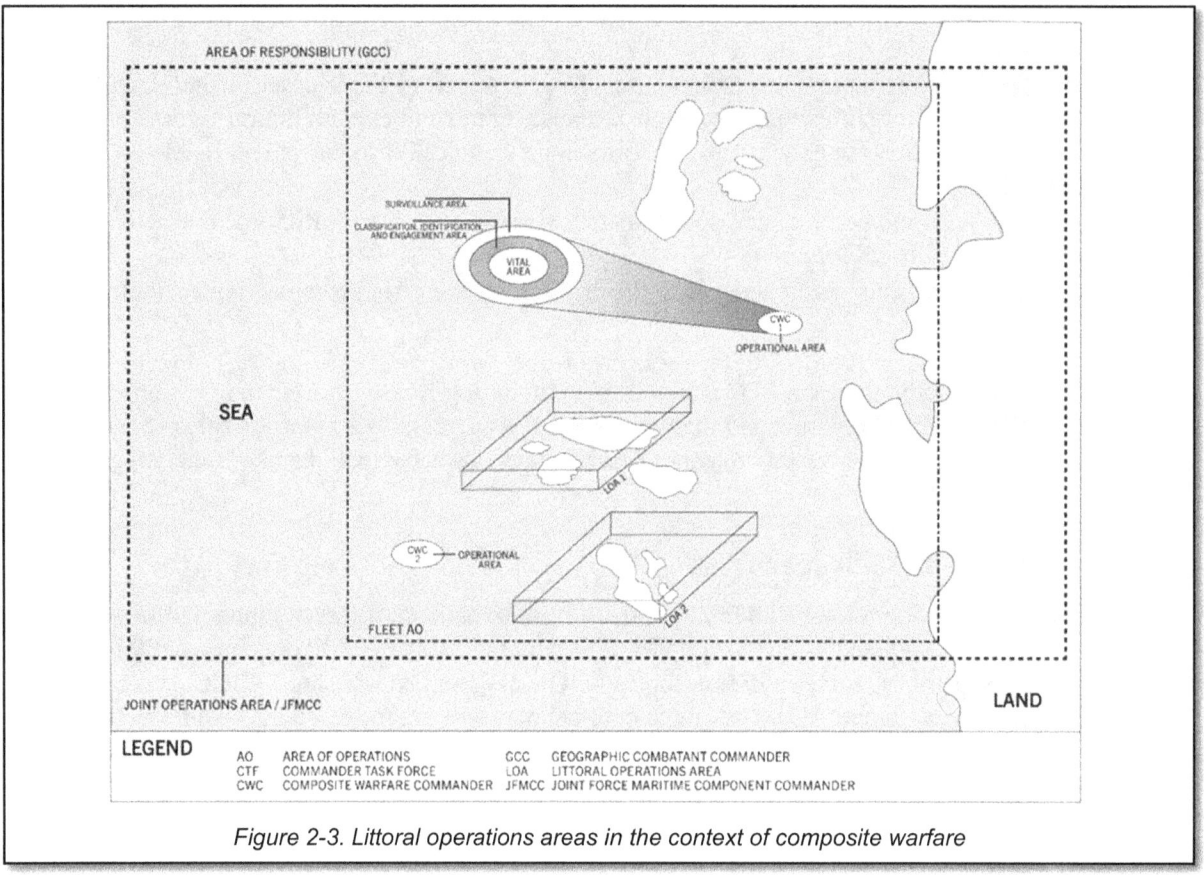

Figure 2-3. Littoral operations areas in the context of composite warfare

Littoral Operations Area. Among the battlespace control measures the JFMCC and fleet commanders may use, this manual proposes for experimentation two forms of a LOA within the maritime AO: the LOA as battlespace and the LOA as a permissive control measure. See figure 2-4 below.

LOA as Battlespace. When the JFC appoints a JFMCC, the JFC will normally designate a maritime AO. The JFMCC may then establish subordinate maneuver space for subordinate elements. The LOA encompassing both landward and seaward littoral terrain may be assigned as subordinate maneuver space. The designation of the LOA as battlespace within the maritime AO is intended to ensure unity of effort and the integration of resources under a fleet or JFMCC commander to accomplish assigned missions. This could include controlling a maritime chokepoint or controlling portions of the littorals necessary to support the fleet's freedom of maneuver and operational design. The designation of the LOA as battlespace assigned to one subordinate commander should not exclude other naval force actions within the LOA such as transit or coordination of fires; these actions simply require the coordination and approval of the commander within the LOA. The authority to designate the LOA within the maritime AO may be retained at a level as high as the JFMCC.

LOA as a Control Measure. Within the established battlespace of the maritime AO and the subordinate maneuver space, the LOA may be a control measure. As a control measure, it could be assigned by the fleet commander to a subordinate commander for positioning of forces, or assigned by a CWC for the EXWC for maneuver of expeditionary forces. This requires the integration of the CWC's resources across specified domains and within the limits of the LOA.

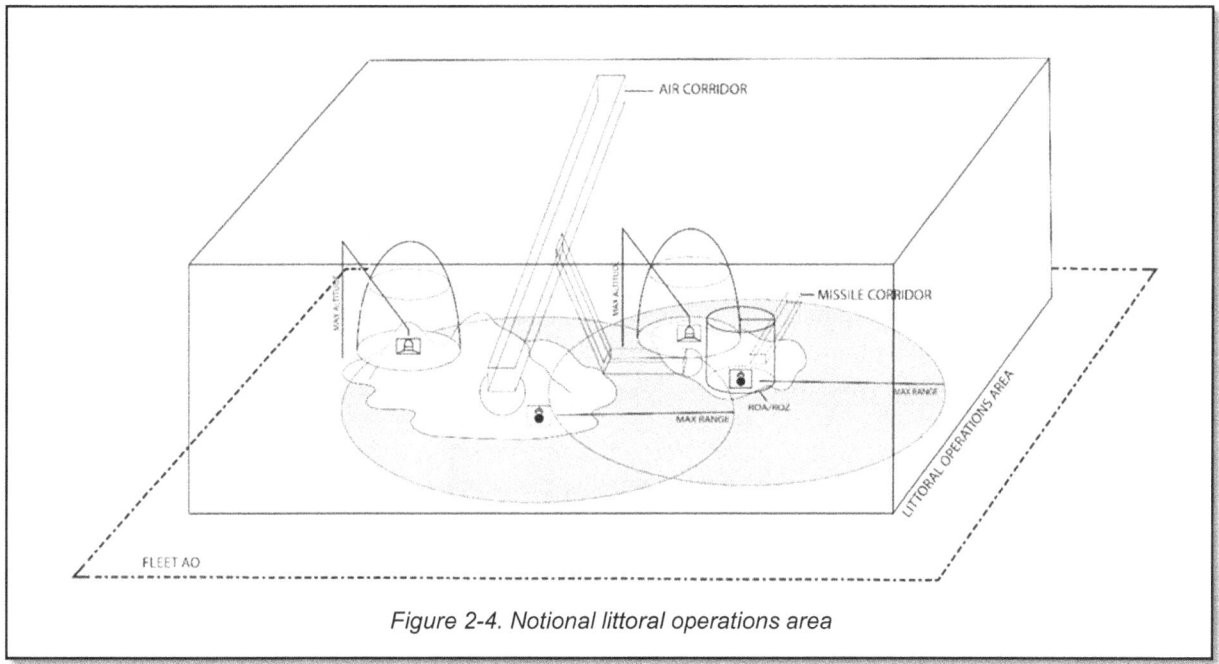

Figure 2-4. Notional littoral operations area

Considerations for LOA Planning and Development. The LOA is a multidomain control measure. Within composite warfare, the LOA enables a commander designated as the EXWC employing littoral forces to mass the combined resources of the CWC within the LOA. The CWC, appointed by the OTC, may in turn appoint functional or subordinate warfare commanders. The EXWC, acting within the limits of the LOA, is effectively a subordinate warfare commander responsible for integrating the resources of the task group to achieve a specific outcome within the three-dimensional limits of the LOA. Simultaneously, the EXWC remains responsive to the requirements of hierarchically adjacent warfare commanders.

Within the LOA, each unit of action will be assigned a *sector*, which is an area designated by boundaries within which the unit will operate and for which it is responsible. Units of action may also be assigned *engagement areas* wherein the commander intends to contain and destroy an enemy force with the effects of massed weapons and supporting systems. Forces assigned responsibility for engagement areas must ensure that their internal fire support coordination measures support the requirements of the engagement area.

Naval Battlespace Terminology Related to Afloat Formations. Three doctrinal terms associated with battlespace constructs for operations of composite task organizations at sea, discussed below in further detail, should also be understood by forces conducting EABO: *surveillance area* (SA); *classification, identification, and engagement area* (CIEA); and *vital area* (VA). The CWC defines the task force's protected asset(s), and warfare commanders such as the sea combat commander (SCC) or AMDC define the ranges associated with these battlespace constructs. Littoral forces must consider the requirements of these battlespace constructs when positioning assets. Through experimentation, they must also explore how landward forces might contribute to operations in the following areas defined within composite warfare.

- Surveillance Area. In surface warfare, a SA encompasses the OE that extends out to a range that equals the force's ability to conduct a systematic observation of a surface area to detect vessels of military concern. The dimensions of the SA are a function of strike group surveillance capabilities, sensors, and available theater and national assets.[12]

- Classification, Identification, and Engagement Area. In maritime operations, a CIEA describes the area within the SA and surrounding the VA(s) in which all objects detected are classified, identified, and monitored. Within the CIEA, friendly forces maintain the capability to escort, cover, or engage. The goal is not to destroy all contacts in the CIEA, but rather to make decisions about actions necessary to mitigate the risk each contact poses. The CIEA typically extends from the outer edge of the VA to the outer edge of where surface forces effectively monitor the OE. It is a function of friendly force assets/capabilities and reaction time, threat speed, the warfare commander's desired decision time, and the size of the VA.[13]
- Vital Area. A VA is a designated area or installation defended by air defense units. The VA typically extends from the center of a defended asset to a distance equal to or greater than the expected threat's weapons release range. The intent is to engage threats prior to them breaching the perimeter of the VA. The size of the VA is a function of the anticipated threat. In some operating environments, such as the littorals, engaging threats prior to their breaching the VA is not possible because operations are required within the weapons-release range of potential threats. Preplanned responses shall include measures for engaging contacts initially detected within, rather than outside, the VA.

Note: Potential exists for multiple organizations to conduct operations within a JFMCC's AO. To ensure unity of command and unity of effort, the JFMCC should ensure common processes and procedures exist for the shifting of tracking across organizational seams.[14]

Control Measures. The Naval Service is used to coordinating operations through their battlespace in three dimensions. Control measures coordinating maneuver, fires, and airspace are critical for managing the battle, providing operational flexibility and minimizing risk. The littoral force will plan and coordinate control measures that cover all domains and enable integration with the larger joint fight. This system must be clearly communicated at all times but also include processes to allow for responsive action in a communication degraded/denied environment.

Existing doctrinal terms, symbols, and naming conventions will be used, as appropriate, when designating control measures for surface and air movement and maneuver—whether seaward or landward—in conjunction with EABO.

Littoral Transition Point (LTP). These are locations where forces conducting surface littoral maneuver will shift between waterborne and overland movement in either direction. Normally, forces conducting EABO will preplan multiple LTPs and avoid repeated use of the same point in order to reduce the likelihood of detection and targeting. Arabic numerals will be used to number LTPs.

2.7 COMMAND AND CONTROL

After receipt of the initiating directive, mission analysis, and task organization of forces, the commander must prepare the staff and subordinate elements to function within the assigned command structure. It should be expected that command relationships will remain dynamic and may rapidly shift based on operational requirements. Commanders must ensure that their staffs are prepared to execute transitions between command structures. For example, the commander and staff may initially be subordinate to a

---

[12] OPNAV, *Composite Warfare*, NWP 3-56.
[13] OPNAV, *Composite Warfare*, NWP 3-56.
[14] OPNAV, *Composite Warfare*, NWP 3-56.

MEF within the MAGTF construct and then become a task element subordinate to a task group within the CWC construct.

Supporting Relationships/Support Situations. While a commander may establish and direct the nature of support between two subordinate commanders, this authority is less commonly employed by naval forces. Instead, naval commanders prefer to establish supporting situations. Supporting situations establish supporting and supported commanders, specify the level of integration with the supported commander, and do not modify existing command relationships.

The Marine Corps provides for the establishment of supporting and supported commander relationships; similarly, the Navy provides for a common commander to direct the nature of support between subordinates without establishing or modifying command relationships. This direction is known as a *support situation* (SUPSIT). In a supporting situation, the respective commanders are designated as either the supporting or supported commander. The Navy recognizes three forms of SUPSITs, as figure 2-5 below illustrates, differentiated by levels of integration and the discretion granted to the supporting commander. The commander directing the support operations must specify the type of SUPSIT between the subordinate commanders. SUPSITs establish collaborative relationships between subordinate commanders without the necessity to modify command relationships.[15] Forces conducting EABO execute under this SUPSIT framework.[16]

SUPPORT SITUATION	ADVANTAGES	DISADVANTAGES
ALPHA (INTEGRATED) TWO OR MORE FORCES JOIN INTO ONE FORCE	UNITY OF COMMAND, EFFORT, AND FOCUS INTEGRATED PLANNING AND SYNCHRONIZED EXECUTION MASSED FORCES FOR MISSION EXECUTION LESS DUPLICATION OF EFFORT, BETTER CONSERVATION OF ASSETS ENHANCED COORDINATION OF ASSET APPOINTMENT BETTER RESOLUTION OF COMPETING TASKS AND PRIORITIES ASSIGNED TO MULTI-MISSION PLATFORMS	REQUIRES MERGER OF TWO OR MORE SEPARATE ORGANIZATIONS INCREASED LEVEL OF EFFORT/C2 REQUIREMENTS FOR THE OTC AND STAFF POTENTIAL LOSS OF FOCUS/TEMPO OF OPERATIONS WHILE TRANSITIONING TO THE NEW COMMAND STRUCTURE
BRAVO (COORDINATED) TWO OR MORE FORCES REMAIN SEPARATE; SINGLE OTC DIRECTS THE TACTICAL OPERATIONS OF ALL FORCES.	COORDINATED TACTICAL OPERATIONS BETWEEN NON JOINED FORCES CENTRALIZED PLANNING, DECENTRALIZED EXECUTION	DECREASED UNITY OF COMMAND, INCREASED LEVEL OF EFFORT/C2 REQUIREMENTS FOR THE OTC AND STAFF HARDER TO AVOID MUTUAL INTERFERENCE OR ELIMINATE REDUNDANT EFFORTS THAN SUPSIT ALPHA MORE DIFFICULT TO DEVELOP SHARED SITUATIONAL AWARENESS BETWEEN STRIKE GROUPS THAN SUPSIT ALPHA POTENTIAL SLOWER DECISION MAKING PROCESS THAN SUPSIT ALPHA
CHARLIE (DISCRETION) TWO OR MORE TASK ORGANIZATION COMMANDERS COORDINATE ACTIONS.	NO CHANGE IN COMMAND STRUCTURE OR C2 REQUIRED TASK ORGANIZATION COMMANDERS CONTINUE TO FOCUS ON THEIR RESPECTIVE MISSION(S) TASK(S)	HARDER TO COORDINATE AVOIDANCE OF MUTUAL INTERFERENCE AND ELIMINATION OF REDUNDANT EFFORTS THAN SUPSIT'S ALPHA OR BRAVO SUPPORT CHANGE REQUIREMENTS MAY TAKE LONGER TO OBTAIN REQUIRES SIGNIFICANT COORDINATION TO OBTAIN UNITY OF EFFORT

Figure 2-5. Navy supporting situations

As stated in subsection 2.6.5, littoral forces may operate under the CWC of an afloat task group or as a separate task group with its own CWC. Depending on the situation, littoral forces may be fully integrated into a CSG or ESG CWC. For example, a CSG may be given tactical control (TACON) of littoral forces to perform a coordinated strike against an enemy SAG. Establishing a separate CWC is likely the preferred choice for longer duration operations to avoid tying the seagoing elements of the fleet, such as

---

[15] OPNAV, *Composite Warfare*, NWP 3-56.
[16] OPNAV, *Composite Warfare*, NWP 3-56.

CSGs, ESGs, and SAGs, to relatively confined operating areas. If a separate CWC is established, the littoral force may be tasked with providing support to other naval task groups or, conversely, receive support from such groups. In such cases, SUPSITs Bravo and Charlie are the most likely options, since the littoral force cannot "join or integrate with" underway task groups.

CHAPTER 3

INTELLIGENCE OPERATIONS

3.1 GENERAL

Providing a positive aim for succeeding in campaigning helps to align actions taken toward reaching intermediate goals. In both current and future operations, focus must be placed on the competitor to develop an understanding of their systems and exploitable vulnerabilities. Likewise, commanders must continually refine models to evaluate the reactions of competitors and partners alike. Further, actions must be taken to conduct self-assessment for awareness of internal vulnerabilities that may be exploited. As such, operations are designed to stimulate known aspects of an adversary system. Intelligence collection planning efforts are purposefully integrated to observe and measure anticipated and unanticipated adversary systems and network deviations.

The naval character of EABO requires integrated intelligence operations to support planning, execution and assessment. The foundation of intelligence support to EABO is battlespace awareness, achieved through continued sensing of the maritime environment to identify baseline conditions and deviations across the competition continuum. There are several challenges inherent to this effort, including the complex nature of the maritime domain, the dynamic nature of the competition continuum, and the finite number of capabilities and resources with which to execute intelligence operations. Establishing maritime domain awareness (MDA) and an OE baseline are necessary steps to meet these challenges. Establishing and exercising partnerships with joint, interagency and multinational partners can provide additional intelligence capabilities and capacities. Local law enforcement agencies are likely good sources in their own region, for example. These partnerships also serve as the building blocks to help plan and execute subsequent intelligence operations.

MDA refers to the understanding of anything associated with the maritime domain that could impact the security, safety, economy, or environment of a nation. Obtaining and maintaining accurate MDA is a key enabler of an active and layered maritime defense in depth. It facilitates expeditious and precise actions by the JFMCC and subordinate commanders, and it enables effective integration with joint force operations.[17] Day-to-day littoral force activities during campaigning will establish an operational baseline that informs the littoral force commander's future decisions. Ongoing sensing both enables awareness of actual or potential escalatory and de-escalatory baseline deviations, and it facilitates future planning and execution. The primary method for information sharing, gaining situational awareness, and supporting collaborative planning in the maritime domain is through development and maintenance of a maritime common operating picture (COP) at both the tactical and operational level. EABs will play a critical role by feeding the CWC COP tactical information and the maritime operations center (MOC) COP operational information within the WEZ. Networking EABs into the maritime COP can present useful data in a form that supports a wide range of planning, decision, execution, and assessment requirements. This data can also support combatant commander (CCDR) requirements to achieve an area of responsibility (AOR) -wide, single integrated COP.

Integrated intelligence operations in support of EABO require the judicious application of intelligence capabilities and resources, organic, theater, and national, to meet the challenges of the maritime domain's scale and complexity, as well as those of the competition continuum. This should include open-source

---

[17] JCS, *Joint Maritime Operations*, JP 3-32.

information and contributions from allies, partners, and indigenous populations where feasible. The ability to provide intelligence to and leverage intelligence from the greater Intelligence Community (IC) is also fundamental to integrating the EABO intelligence effort. Littoral forces must execute daily intelligence operations during campaigning to achieve campaign objectives in competition, prevent and deter crisis, and set conditions for success in conflict. This chapter outlines the integration points and planning considerations that are necessary to provide effective intelligence support to EABO within a highly contested environment against peer adversaries.

3.2 PURPOSE AND SCOPE

Intelligence operations seek to provide the littoral force with an understanding of how friendly, enemy, and neutral activities across the competition continuum can impact EABO efforts. Intelligence support to EABO requires an evolutionary step in the continuous development of a Navy and Marine Corps intelligence, surveillance, and reconnaissance (ISR) enterprise that must be validated through rigorous experimentation, training, exercises, and operations. The concepts below do not change *what* Marine Corps and Navy intelligence provides from a functional perspective. They propose instead changes to *how* intelligence operations provide awareness and responsiveness to littoral forces in the transition from campaigning to conflict.

3.3 INTELLIGENCE-LED OPERATIONS

Operations during prior conflicts have demonstrated that intelligence collection is the most challenging part of the intelligence and operations cycle. Experience has revealed a need to understand what a potential adversary's collection apparatus can and cannot sense and make sense of, which drives a shift from pursuing "actionable intelligence" alone to conducting "actions to produce intelligence." Actions to produce intelligence are termed "intelligence-led operations." Intelligence-led operations, to include traditional and non-traditional intelligence activities, are crucial to EABO in campaigning because they help develop a baseline picture of the maritime OE. The baseline allows the fleet to identify anomalies, act preemptively to deter or counter the adversary's plans, and prepare for escalation to armed conflict.

While conducting EABO during campaigning, the projection of information, such as key leader engagements, subject matter expert exchanges, bi-lateral exchanges, and other activities provide littoral forces the opportunity for engaging and influencing key actors within the HN and collection and baseline development of the adversary in littoral areas. This baseline should span the entire scope of the littoral environment encompassing key terrain; adversary presence and activity; areas, structures, capabilities, organizations, people, and events (ASCOPE) considerations; political, military, economic, social, infrastructure, and information (PMESII) considerations; and friendly, adversary, and neutral activities in the IE. The need to baseline areas during campaigning should be an important factor in the prioritization of exercises, exchanges, and EABO activities.

Locating and targeting the adversary's concealed sensors and anti-access/area denial weapons require effort beyond waiting and watching with networked sensors. EABO missions provide littoral forces an opportunity to employ organic ISR and maneuver elements to draw adversary forces from cover, learn their tactics, and baseline their operations. Having established the baseline during campaigning, littoral forces can create a collection plan and conduct activities within the OE and IE to capture responses by adversary forces, local populations, and other relevant actors and target audiences. EABO provide LFCs the opportunity to "raise the noise floor" and generate adversary activities in response. EABO intelligence operations should seek to collect, analyze, and distribute relevant and timely intelligence products based on these adversary actions.

In sum, intelligence-led operations in support of EABO seek to generate adversary activity against which littoral forces will plan and execute future missions across the competition continuum. The following paragraphs explain several efforts that are critical to this process.

3.4 NAVAL AND JOINT FORCE INTEGRATION

The primary objectives of Naval Service intelligence - to provide accurate, timely, and relevant knowledge about both the enemy and the environment[18] - still apply in EABO. These objectives become broader in scope when supporting battlespace awareness about the greater joint environment within which EABO will be executed. Along with these primary objectives, joint intelligence operations are responsible for countering adversary deception and surprise, supporting friendly deception efforts, and assessing the effectiveness of operations.[19] Littoral force intelligence operations enable EABO by utilizing organic intelligence resources and capabilities to support and extend the JFMCC's maritime sensor network. As part of a naval force, littoral force intelligence operations not only provide the awareness necessary for littoral forces to execute effectively, but also support joint requirements for JFMCC and JFC decisions.

Close coordination with joint, coalition, HN, and national intelligence organizations early in planning is essential to align platform and sensor employment plans. This optimizes ISR and associated processing, exploitation, and dissemination systems throughout the joint force. The complexity of operating in the maritime domain requires a baseline of organic intelligence collection assets, augmented by joint forces and capabilities allocated to maintain MDA and to succeed in military operations.[20] The collective JFMCC effort to provide the maritime perspective on the OE is crucial to attaining joint force objectives. To support this effort, the JFMCC advocates for the use of other component and national level assets to provide the optimum support to maritime operations. The littoral force's role in providing intelligence support to joint operations is essential to the allocation of limited intelligence resources and the planning and execution of joint force operations.

Special operations forces (SOF) integration is an important piece of joint force integration. SOF provide perhaps their greatest value to EABO and the littoral force through support to intelligence operations. Early in cooperation and competition, SOF-unique authorities, relationships, and capabilities help to answer priority information requirements and illuminate the operating environment. Operating forward with a small, tailored footprint and typically working with other agency, coalition, and indigenous partners, SOF connect military activities to other intelligence networks. These preparation of the environment activities inform the JFMCC effort and commander's decision making, while also operating within the adversary's decision-making cycle.

While the littoral force enables SOF operations early in cooperation and conflict, SOF activities and operations enable EABO and the littoral force by setting favorable conditions and preparing and shaping the environment to meet commander's intent. Marine Forces Special Operations Command's (MARFORSOC) value as a strategic-shaping force will support wider intelligence collection by connecting the littoral force to US Special Operations Command, intelligence agencies, and coalition forces, assets, and capabilities. Likewise, the littoral force and EABO both enable and are supported by US Space Command and US Cyber Command.

In the context of this tentative manual, integration of intelligence with a numbered fleet includes collecting against the fleet's information requirements and sharing targeting data, collections tasking,

---

[18] Headquarters, US Marine Corps, *Intelligence*, MCDP 2 (Washington, DC: US Marine Corps, 2018).
[19] Joint Chiefs of Staff, *Joint Intelligence*, JP 2-0 (Washington, DC: US Department of Defense, 2013).
[20] JCS, *Joint Maritime Operations*, JP 3-32.

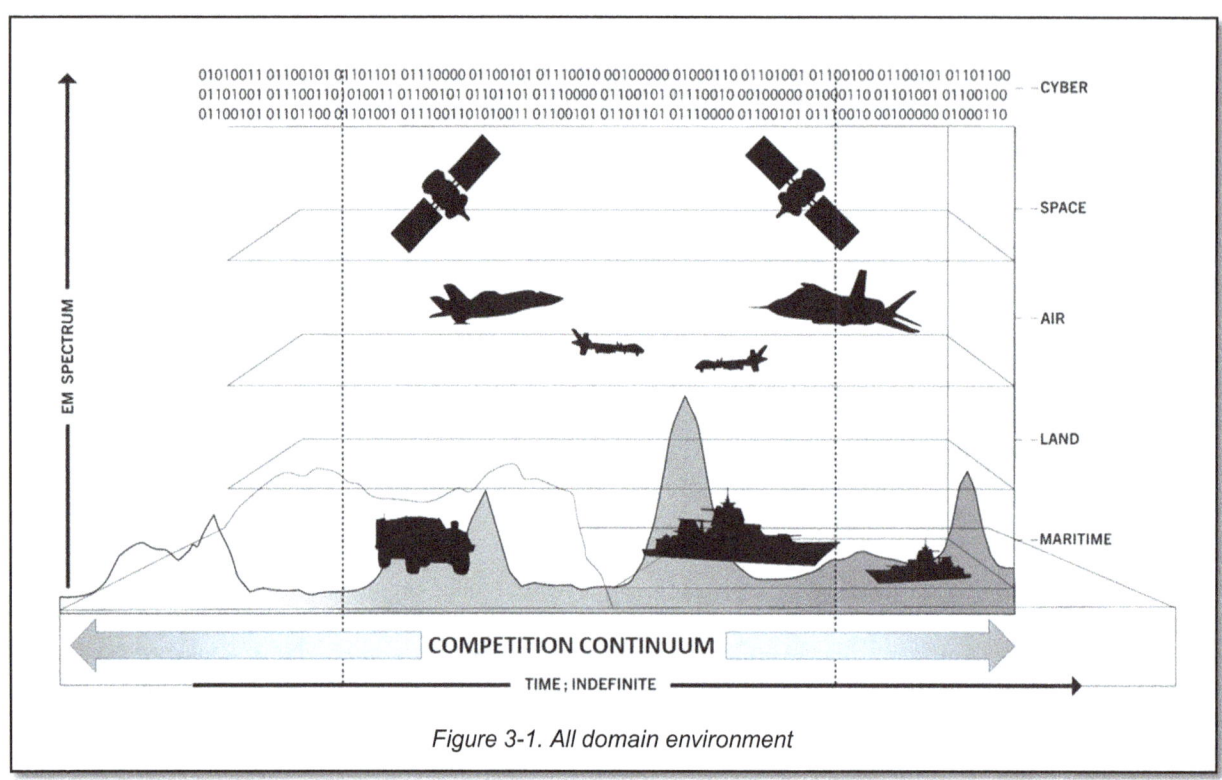

Figure 3-1. All domain environment

battle-damage assessments, and tactical running intelligence estimates. Similarly, theater integration may consist of coordination of the same with the standing joint information center and joint intelligence operations center, as well as joint intelligence coordination with other service components. Integration with the IC may consist of validating and using national intelligence estimates, accessing and contributing to real-time intelligence broadcast feeds and collection lines, and satisfying national collection requirements. Finally, integration with allies and partners may consist of gathering local atmospherics, validating access assumptions, and leveraging their collection platforms and data.

3.5 OPERATIONAL ENVIRONMENT

The OE is "the aggregate of the conditions, circumstances, and influences that affect the employment of capabilities and bear on the decisions of the commander."[21] It includes the land, maritime, air, space, and cyberspace domains, the EMS, and the information environment, as shown in figure 3-1. Intelligence operations must focus on the littoral areas across the domains to support EABO planning. Understanding the littoral OE enables EABO problem framing, determining battlespace geometries, identifying resource shortfalls, identifying critical information requirements, and developing assessment measures. All of these aspects must be understood through the lens of campaigning to develop an OE baseline for future operations.

The principles and steps of the joint intelligence preparation of the operational environment (JIPOE) process are no different for EABO than for other types of operations. However, to support EABO effectively, the process must integrate multidomain naval considerations in the conduct of operations throughout the maritime environment.

---

[21] Joint Chiefs of Staff, *DOD Dictionary of Military and Associated Terms* (Washington, DC: US Department of Defense, 2022).

3.5.1 Contested Environment

EABO are executed in contested maritime environments, which encompass both uncertain and hostile environments that will vary by time and space. Whether those environments are influenced or controlled by state or nonstate actors, littoral forces face a variety of direct and indirect multidomain threats capable of limiting, interrupting, degrading, or denying their ability to attain objectives. This contested space presents a number of challenges for intelligence operations, not least of which is adherence to EABO characteristics.[22] A significant challenge is understanding how to plan and execute operational activities to facilitate intelligence collection. In a contested environment where the adversary seeks to inhibit the littoral force's freedom of action, littoral force activities may elicit an adversary response that exposes a capability or vulnerability that the littoral force can exploit for a future competitive or combat advantage. This idea of "intelligence led operations" is explained in further detail below in section 3.6. As always, it requires close coordination and synchronization of effort between intelligence and operational planners. Littoral forces plan and execute intelligence operations to facilitate cooperation with partners and allies in competition and create operational flexibility when escalation leads to armed conflict. Establishing an OE baseline and maintaining MDA require persistent awareness and analysis of the OE to determine whether actions therein constitute an escalation of intentions beyond competition. Littoral forces must build awareness beyond the traditional view of the physical environment to include IE awareness. Understanding the adversary's IE activities during campaigning can directly impact the littoral force's ability to operate effectively across the competition continuum.

3.5.2 The Information Environment

The IE is the aggregate of social, cultural, linguistic, psychological, technical and physical factors that affect how humans and automated systems derive meaning from, act upon, and are impacted by information, including the individuals, organizations, and systems that collect, process, disseminate, or use information.[23] This definition expands the physical scope of potential variables that are capable of directly and indirectly impacting the OE. The Marine Corps uses the term IE to refer to the global competitive space that spans the warfighting domains, where all operations depend on information. It includes information itself and all relevant social, cultural, psychological, technical, and physical factors that affect the employment of forces and bear on commanders' decision making.[24] Actions in the IE across the world may potentially inhibit the littoral force's ability to conduct EABO in a designated littoral OE. When littoral forces plan and execute operations, they must understand the real and potential impacts from and to adversary, friendly, and neutral perspectives and how these impacts extend beyond the battlespace. As part of the integrated JIPOE process, littoral forces should analyze physical, human, and informational aspects of both the OE and IE.

Physical aspects are the natural and man-made environmental characteristics that enhance or inhibit communication. Human aspects are the *interactions* among humans, and between humans and the environment, which influence decisions. Informational aspects of the OE reflect the ways that individuals, groups, and human networks communicate and exchange information. Interaction among these aspects within littoral areas adds to the complexity of EABO planning and execution but must be accounted for to enable mission success.

Intelligence operations provide awareness of the adversary's capabilities in the IE. Use of the EMS, cyberspace, and space is critical to both the protection of friendly networks and the identification of adversary networks, systems, and information to attack and exploit. Systems and network analysis can

---

[22] EABO characteristics include stand-in engagement, low signature, etc. Refer to subsection 1.4, "Characteristics of Expeditionary Advanced Base Operations" for the complete list.
[23] JCS, *Information in Joint Operations*, JP 3-04.
[24] Headquarters, US Marine Corps, *Information*, MCDP 8 (Washington, DC: US Marine Corps, 2022).

support the identification of target audiences, relevant actors, and key influencers and decision makers to inform the planning and execution of influence and deception operations.

3.5.3 The Littoral Environment

The littoral environment (figure 3-2) includes a complex combination of seaward and landward physical areas and international legal considerations within the greater maritime environment. Joint maritime operations occur in blue water (high seas and open oceans), green water (coastal waters, ports, and harbors), and brown water environments (navigable rivers, lakes, bays, and estuaries), and in littoral landward areas.[25] Each area presents unique characteristics that intelligence efforts must identify to provide LFCs the awareness necessary to make informed operational decisions. EABO are relevant to the entire maritime domain but are optimized for the littorals. Littoral forces must recognize the littorals as one contiguous area defined by two segments, the seaward and the landward. The seaward segment contains the area from the open ocean to the shore, which littoral forces must control to support operations ashore. The landward segment includes the inland areas, which littoral forces seek to support and defend directly from the sea. A JIPOE for a littoral OE must account for adversary capabilities and limitations across both segments, including how their forces operate within and between each in response to potential threats, and aspects of the physical terrain (surface and subsurface) that inform maneuver from the sea to land-based objectives.

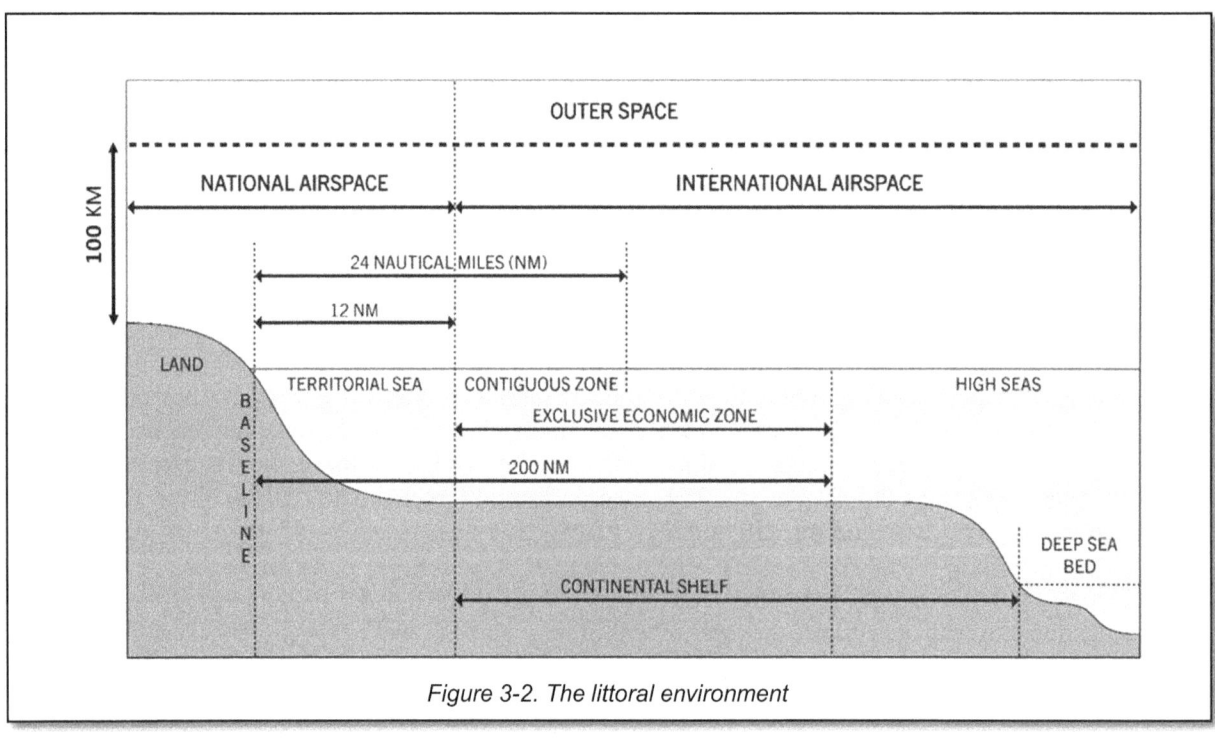

Figure 3-2. The littoral environment

The cross-domain physical features in littoral areas create key maritime terrain relative to adversaries and allies throughout the world. EABO enable naval forces to shape actions and influence events to support sea control and sea denial through their ability to operate effectively in these areas. Littoral force commanders must understand the elements that make the littoral areas key terrain for both friendly and adversary forces. An integrated JIPOE process establishes a baseline understanding of the littoral OE in which littoral forces will execute a variety of missions. Understanding the unique physical characteristics

---

[25] JCS, *Joint Maritime Operations*, JP 3-32.

and impacts of the littoral environment—and adversary capabilities and COAs within that environment—facilitates EABO planning.

3.5.4 Network Analysis and Civil Considerations (PMESII/ASCOPE)

Variables beyond the physical aspects of the OE also impact operations. EABO require an understanding of the unique diplomatic, informational, military, economic, and legal aspects that are specific to the maritime domain. Littoral environments comprise complex intersections of physical terrain, civil and commercial infrastructure, international and state laws, and cultural and social dynamics. In order to account for these variables, littoral forces should conduct an analysis of relevant networks and civil considerations as part of the integrated JIPOE process to support the littoral OE assessment.

Network engagement consists of interactions with friendly, neutral, and threat networks, conducted continuously and simultaneously at the tactical, operational, and strategic levels, to help achieve the commander's objectives within an operational area. Network analysis views an area through the PMESII networks and cells that are present, then analyzes how those elements interact and impact the OE. Understanding the composition and interaction of networks and cells relevant to the littoral OE in campaigning will help littoral forces determine how best to set conditions to achieve objectives and prepare for potential escalations to armed conflict. An accurate understanding of the OE requires cross-functional participation by staff elements of the joint force and collaboration with various intelligence organizations, US government agencies, and nongovernmental subject matter experts.[26] This integration and coordination ensures a common understanding of the littoral OE across echelons of command. Planners can find a more comprehensive discussion of network engagement and analysis in *Joint Countering Threat Networks, JP 3-25* and *Network Engagement: Targeting and Engaging Networks, MCTP 3-02A*.

Understanding civil considerations in the littoral OE is critical to effective littoral force actions in competition. Littoral force shaping actions in campaigning seek to set the conditions to maximize freedom of action. Building a framework of civil considerations establishes a baseline awareness of the ASCOPE that are specific to the littoral OE where littoral forces are likely to conduct EABO. The ASCOPE framework provides LFCs an understanding of the current state of critical conditions that must be met to enable future EABO missions. From the initial ASCOPE assessment, littoral forces should continue to plan and execute operations, activities, and investments to influence the conditions and shape the environment to enable EABO execution. Planners can find a more detailed discussion of the ASCOPE framework in *Civil-Military Operations, JP 3-57*.

In the continuous preparation for transition from competition to conflict, the Marine Corps Intelligence, Surveillance, and Reconnaissance Enterprise (MCISRE) analyzes target systems through an open-ended process called target systems analysis (TSA), utilizing all-source, fused intelligence to choose potential targets that, when engaged, are most likely to create desired effects that contribute to achieving the LFC's objectives. TSA is a process of identifying, describing, and evaluating the composition of an adversary target system and its components to determine its various functions, capabilities, requirements, and vulnerabilities. TSA is further refined to exploit target system vulnerabilities (e.g., target development at the entity level) that weaken the adversary's ability to engage successfully during competition or hostile operations. This process can be adapted for employment during operations in the campaigning phase to identify targets in the IE, such as PMESII systems or subsystems and components of the ASCOPE

---

[26] Joint Chiefs of Staff, *Joint Countering Threat Networks*, JP 3-25 (Washington, DC: US Department of Defense, 2016).

framework. Planners and analysts can find a more detailed discussion of target systems analysis in *Joint Targeting School Student Guide*.[27]

3.6 INTEGRATED NAVAL INTELLIGENCE PROCESS

Littoral forces must integrate Marine and Navy intelligence efforts at every level possible, to include the littoral force staff, littoral force intelligence enablers, and at higher echelons of command. This integration should occur across multiple lines of effort:
- Employing integrated systems
- Ensuring system interoperability
- Integrating cross-domain solutions (i.e. JWICS to SIPRNET, SIPRNET to Coalition networks, etc.)
- Training and exercising littoral force ranks for cross-functional proficiency
- Ensuring interdisciplinary intelligence proficiency
- Synchronizing boards, centers, cells, and working groups at echelon

Cross-functional proficiency seeks incorporation of intelligence operations with activities of other warfighting functions. Interdisciplinary proficiency helps avoid overreliance on one type of intelligence to support operational planning and execution. Littoral forces should seek both forms of proficiency to maximize the capacity of their intelligence efforts across Navy and Marine Corps capabilities and resources.

Littoral forces capitalize on naval integration as a fleet asset operating in the joint and coalition environment through coordination, collaboration, and cooperation. Marine ISR capabilities resident in the littoral force may be employed to answer higher-echelon requirements. For example, littoral forces can fuse intelligence efforts at the tactical level—maritime common operational picture, analytic products and assessments, and IE running estimate—to provide OE awareness to higher echelons and the greater IC. In the larger joint environment, the LFC is able to employ and integrate higher naval and joint assets and capabilities to enhance the intelligence fusion. Working with JFMCC and combatant command J-2 resources and collection processes, the littoral force extends the C5ISRT as a stand-in force.

Intelligence integration above the littoral force level must occur at intelligence sections of the numbered fleets and MEF headquarters levels. Littoral force-generated intelligence should be fused with fleet and MEF intelligence to provide persistent awareness and broader understanding of the joint OE for other littoral forces and adjacent Navy and Marine Corps units throughout the theater. This highlights the push-pull nature of intelligence operations that provides operational awareness across a large geographical area.

3.6.1 Activity-Based Intelligence

Activity-based intelligence (ABI) is an analytic methodology that littoral forces can employ to fully leverage the large volume of data collected by the joint force. ABI can support EABO by rapidly integrating relevant data for area-specific assessments and indications and warning. A challenge to ABI is determining to what extent naval intelligence analysts can execute this methodology at the tactical level to support EABO.

In traditional intelligence analysis, an all-source analyst combines the specialized reporting from various intelligence disciplines into fused intelligence products and assessments. In ABI, collected data is integrated before it is analyzed by a specialist from the particular intelligence discipline that was

---

[27] Joint Targeting School, "Joint Targeting School Student Guide" (student guide, 1 March 2017), https://www.jcs.mil/Portals/36/Documents/Doctrine/training/jts/jts_studentguide.pdf?ver=2017-12-29-171316-067.

responsible for the initial collection. ABI is designed to maximize the power and indicators within big data from multidomain sensors and sources. ABI seeks to rapidly integrate the data and indicators from activities, events, and interactions of actors and systems within the OE and IE to identify and characterize relevant patterns and anomalies, thus creating decision advantage for commanders.

Littoral forces must test and exercise ABI in C2 degraded training environments to understand the constraints and limitations associated with employing this method at the tactical level. ABI requires access to relevant and timely reporting to understand the most operationally significant factors impacting the OE. This may be difficult to execute in the contested maritime environment. Understanding and training with the procedural limitations can generate analytic tactics, techniques, and procedures (TTPs) that identify ABI requirements for tactical intelligence enablers and those that require reach-back support.

ABI is sequence neutral between archived and current data and between incidental collection and planned missions for targeted collections. Once the geo-referenced report is identified, a holistic, all-data analysis generates a multisource product with historical context. Leveraging ABI will enhance littoral force abilities to task sensors dynamically, provide force protection, develop target intelligence, and gain and maintain effective battlespace awareness. Fully leveraging ABI requires littoral forces to obtain support from intelligence sections at numbered fleets and MEFs. This support is necessary to both ensure the greatest possible access to available data and mitigate the potential effects of operating in a degraded communications environment. Achieving this level of integration between forward forces and reach-back support will require deep familiarity between the two. The ABI methodology and effective integration between forward forces and reach-back support should be practiced and exercised at every opportunity to ensure proficiency when operating in a contested maritime environment.

Successful execution of ABI depends on, but is not limited to, the following considerations to support EABO:
- Battlespace awareness: provide persistent surveillance of the landward and seaward portions of the littorals and leverage all sources of information to include ISR collections, open-source intelligence, logistics and transportation data, and meteorological and oceanographic information
- Acquire knowledge of adversary intent and capabilities and an understanding of where, when, and how adversaries operate
- Plan and direct multidisciplinary intelligence, counterintelligence, and reconnaissance operations across all warfighting domains to satisfy the commander's requirements
- Allocate intelligence resources and integrate planning and direction between intelligence, operations, and other staff sections
- Conduct reconnaissance and surveillance of the littoral area
- Convert collected information into forms suitable for further analysis and action
- Conduct technical processing and exploitation of organically collected and joint intelligence in contested maritime environments
- Distinguish abnormal from normal patterns of activities
- Integrate multidomain, geo-referenced data for immediate discovery
- Develop and use artificial intelligence and machine learning to support relevant data processing

3.6.2 Support to the Sensing Enterprise

By nature, EABO extend the naval sensor network. Sensing as an enterprise service is a conceptual vision for naval ISR operations. This approach seeks to shift the sensing paradigm from a specific operational focus to a broader service to support the simultaneous needs of many independent users and to provide more comprehensive situational understanding and battlespace awareness. By combining sensor and multiuser information services (i.e., user software interface), sensing as a service enhances decision advantage for naval and joint forces. The employment of Marine collection platforms during EABO plays a critical role in extending the enterprise-sensing network.

Using integrated maritime intelligence standards, processes, and sensor architectures, the Naval Service seeks the flexible, dynamic, and responsive application of ISR capabilities and resources to the places where they are most needed, unconstrained by traditional organizational and geographic limitations. Ideally, the extensive network of people, systems, data, networks, and services across the Naval Service should expand as the mission dictates to include elements from other services, joint forces, coalition partners, the IC, and other government agencies and organizations.

Successful support to the sensing enterprise depends on, but is not limited to, the following considerations to support EABO:
- Contribute to the naval common intelligence picture
- Share a common tactical operations and intelligence picture that permits cooperative engagements between multidomain platforms
- Conduct ISR mission planning as part of the naval ISR enterprise while aboard ship and ashore in a contested environment
- Issue orders, requests, or tasking to naval and joint intelligence organizations
- Integrate with fleet sensors and leverage artificial intelligence to manage collection requirements and collection operations
- Ingest data from nonintelligence, joint, and coalition sensors
- Plan and distribute multiple layers of intelligence collection data across redundant communication pathways and in accordance with SIGMAN requirements
- Support passive and active battlespace sensing on, above, and below the surface of the OE, as well as in the EMS
- Provide indications and warnings from organic, joint, and combined sensor data while on board ship and ashore
- Integrate with the littoral force's fires architecture and the Naval Operational Architecture to link sensor to shooter for an instantaneous detection-decision-action cycle that reduces kill-chain timelines and enhances fleet lethality through dynamic and responsive targeting and real-time assessments
- Sense beyond the maximum effective range of Marine Corps organic fires to detect, establish positive identification, and derive target-quality location data of adversary naval and proxy forces using organic, naval, and joint sensors for the littoral force and designated CWC(s)

3.6.3 Collections Planning

Collections includes those activities related to the acquisition of data required to satisfy specified information requirements that support planning and operational efforts, including establishing OE and IE baseline development, gaining and maintaining MDA and IE battlespace awareness, informing SIGMAN and force protection, and support to assessments.[28]

Day-to-day littoral force collection activities should support higher echelon collection requirements and set conditions to execute EABO. Maintaining MDA for forces throughout the theater is a persistent requirement. Along with support to theater collection requirements, littoral forces must continually develop a maritime OE baseline of adversary and neutral networks and activities during campaigning to prepare for potential escalations to armed conflict. The critical elements drawn from the integrated JIPOE process allow littoral force staffs and commanders to discern key terrain, adversary disposition and

---

[28] JCS, *Joint Intelligence*, JP 2-0.

capabilities, relevant actors, target audiences, and key decision makers within the littoral force OE against which to focus collection efforts.

Intelligence collections is critical to supporting assessments at all levels of war and across the competition continuum. Strategic and operational assessment efforts concentrate on broad tasks, effects, objectives, and progress toward specified end states.[29] Littoral force collection platforms must use this approach to assess the success or failure of efforts to shape the maritime environment in support of potential EABO. Successful planning and execution of a collections plan depends on, but is not limited to, the following considerations to support EABO:

- Gather intelligence data and information from all intelligence disciplines across all domains to satisfy identified requirements
- Create persistent and resilient sensor operations regardless of time of day or weather conditions
- Visualize coverage of both intelligence-directed and nonintelligence sensors in operations and layer multidomain sensors to fill collection gaps
- Disseminate collection data across multiple communications pathways to ensure forces at the EABs receive force protection and targeting data while also supporting joint targeting efforts
- Perform rapid and accurate combat assessments of littoral and naval targets in the littoral force's area of interest
- Conduct battle damage assessment

3.6.4 Support for Effective Signature Management

SIGMAN is critical to the survivability of Marine forces executing EABO missions within the adversary's WEZ. The ability to alter or limit observable and measurable signatures will preserve and extend the capabilities and proficiency of personnel and systems supporting EABO by making them more difficult to identify and target.

Adversaries employ a variety of collection and sensing platforms to support the identification of friendly force locations, purposes, personnel, and systems. Intelligence support to SIGMAN should focus heavily on these adversary's collection and targeting assets. The aim is to develop an overall threat picture that allows littoral force planners to determine what signatures are most vulnerable to adversary collection and exploitation. This analysis should include sensor type and availability, conduit analysis focused on the pathways from sensor to decision maker, specific types of adversary collection assets focused on areas of friendly activity, and the adversary's understanding of friendly force abilities to conduct specific missions. Consideration must be given to the ability of the local population to observe Marine forces, and post geo-tagged photos, videos, or comments on social media that could pose a threat to EABO. Once collection and targeting capabilities are identified, intelligence and operations planners must collaborate to develop an understanding of adversary's information and decision-making processes.

Conduit and emulative analysis must both be conducted after determining the collection and targeting assets to be analyzed. The challenge for littoral forces is the testing and experimentation necessary to determine the level of detail and understanding that can be attained by tactical intelligence analysts to support analysis of friendly force SIGMAN. To prepare for EABO in C2 degraded or denied environments, littoral forces must exercise and test understanding of the constraints and limitations at the tactical level in a forward maritime environment. These analytic efforts may require reach-back support to the numbered fleet and MEF intelligence sections, but they must also manage the challenge of distributing the relevant intelligence analysis to forward forces to support SIGMAN efforts.

---

[29] JCS, *Joint Intelligence*, JP 2-0.

Conduit analysis is a key supporting activity to operations security and deception planning. It is a systems approach to mapping information or intelligence pathways from sensor to decision maker, which includes cross-cueing, transmit time between nodes, effect of filters, and intelligence fusion and dissemination times. Emulative analysis is a psychological and sociological understanding of biases, perceptions, and predispositions of adversary decision makers, which helps identify how adversaries might act based on information available to them.

Applying conduit analysis to adversary targeting activities leads to kill-chain analysis. Kill chains correlate to the process by which a force can find, fix, track, target, engage, and assess (F2T2EA) a target. This analysis is critical to planning effective SIGMAN for survivability. Littoral forces must understand how the adversary observes friendly actions and targets friendly capabilities. By understanding the adversary process, littoral forces avoid wasting time and resources to manage signatures that are unseen or unimportant to the adversary's process. This chain of dependent activities presents opportunities for disruption, destruction, or defeat through offensive action, as well as through active and passive defensive measures. Through a holistic approach to SIGMAN, intelligence operations can support the littoral force's counter-ISR and deception efforts.

Successful planning and execution of intelligence support to SIGMAN operations depends on, but is not limited to, the following considerations for EABO:
- Support protection activities in the IE
- Measure and monitor own-force electromagnetic signature to enable its management and avoid detection during operations
- Sense chemical, biological, radiologic, nuclear, and explosive signatures
- Conduct sensor-to-shooter, kill-chain (i.e., F2T2EA) analysis
- Determine multidomain indicators and signatures associated with sustainment operations by friendly forces
- Detect adversary surveillance of friendly forces and provide electromagnetic countermeasures for targeting data
- Conduct counterintelligence activities to support critical technology protection

CHAPTER 4

INFORMATION ACTIVITIES IN SUPPORT OF EABO

4.1 GENERAL

Naval forces combine maneuver, fires, and information in a 21st century combined arms philosophy to gain and maximize information advantages over our adversaries. Information advantage is an exploitable condition resulting from one actor's ability to generate, preserve, deny, and project information more effectively than another. EABO are designed to influence strategic, operational, and tactical decisions of friendly forces, allies, and adversaries across the competition continuum, deterring or countering adversary actions and enabling friendly actions. Conversely, adversary actions in the IE originating far outside the physical reach or influence of an EAB may have tangible effects on EABO.

The LFC and staff must integrate the information warfighting function into all operations based on an informed understanding of the OE and identification of the specific effects desired. The commander and the staff must be cognizant that gaining, maintaining, and exploiting information advantages can only be effective if informed by assessment and analysis of the cultural, political, social, and economic factors that influence the objectives and behavior of key actors.

4.2 PURPOSE AND SCOPE

Information activities are actions taken to generate, preserve, project, and deny military information power in order to increase and protect competitive advantage or combat power potential within all domains of the OE. The joint force leverages information across the competition continuum to assure, deter, coerce, and compel relevant actors in pursuit of US national objectives. Littoral forces seek to create and exploit three types of information advantages - systems overmatch, prevailing narrative, and force resiliency - to gain access to adversary command and control networks, build support for US presence, deter adversary aggression, disrupt adversary confidence, expose and counter malign behavior, and protect and defend littoral force C2 and ISR networks.

This chapter serves as a tool for those planning and executing information in support of EABO. It describes the information environment, discusses application of the four functions of information (generate, preserve, deny, and project), information capabilities in the context of EABO, and provides key planning considerations and tasks.

4.3 INFORMATION ENVIRONMENT BASICS

4.3.1 Adversary Activities in the Information Environment

Adversary activities in the IE may seek to frustrate littoral forces by gaining their own information advantages to disrupt and deny US power projection. With respect to systems overmatch, adversaries opposing EABO may employ technical means through cyberspace, space, and the EMS to target, attack, and exploit friendly forces' ability to gather intelligence, understand the situation, and command and control forces. They may also seek the prevailing narrative to foment mistrust, discord, and unrest among the population in the vicinity of EABs, among the forces conducting EABO, and among the US domestic audience. Given the nature of the modern IE, EABs must assume they will be under constant observation. Adversaries will attempt to sway public opinion locally and globally against US forces and the United

States to undermine our alliances, our access, basing and overflight rights, and the will of the American public.

4.3.2 Military Information Advantage

An expanded view of the military instrument of national power includes two mutually reinforcing elements—*physical combat power* and *military information power*. Gaining advantage in these elements is a constant effort applicable across the competition continuum. The ability to manipulate, deny, or destroy the information required by adversaries for the basic functioning of military operations provides significant advantage. Fostering favorable public opinion also generates competitive advantage. Opposing forces, in competition below armed conflict and armed conflict, are in a continuous struggle to gain and maintain these combat power advantages. The essence of military information advantage is the ability to exert one's will or influence over an opponent through the *generation*, *preservation*, *denial*, or *projection* of information. These actions are organized, coordinated, and executed through physical actions conducted in all domains—all integrated and synchronized under the information warfighting function.

4.4 INFORMATION WARFIGHTING FUNCTION

The information warfighting function involves the management and application of information, including its deliberate integration and synchronization with other warfighting functions, to support the planning and execution of operations. Marines use information as a warfighting function to create and leverage information advantages. EABO are intended to alter the behavior of the opponent by communicating messages of credible lethality and demonstrating the resolve to use it. Therefore, EABO are fundamentally an application of the functions of information.

Information is an integral component of broader naval operations. Listed below are seven applications of the information function designed to increase, maintain, or exploit potential competitive or combat power in all domains. These tasks, explained in greater detail below, equip littoral forces with a variety of ways and means to potentially impact strategic outcomes.

4.4.1 Assure Command and Control and Critical Systems

The success of joint operations, including EABO, in all domains requires assured access to trusted information. This requirement preserves the integrity of C2 systems and contributes to decision superiority. The opponent will seek to manipulate, disrupt, or destroy the information within friendly C2, ISR, and weapon systems. Their purpose will be to obfuscate, confuse, and shatter friendly cohesion and deny EABO forces the ability to function and fight. Therefore, forces engaged in EABO must take aggressive offensive and defensive actions to preserve the integrity of friendly information, which includes the fundamentals of both operations security and SIGMAN.

Principles for assuring C2 and critical systems when conducting other types of operations in austere environments apply equally to EABO. Littoral forces must employ systems with small physical footprints capable of low-signature emanations, especially in cyberspace and across the EMS to neutralize or delay the adversary's targeting cycle, thereby creating a time advantage relative to the adversary.

EABO are conducted in contested, degraded, and denied environments to execute specific missions in support of naval and joint operations. Littoral forces, in close coordination with Navy information warfare and Marine Corps Information Maneuver Forces support, must monitor in near-real-time potential threats and vulnerabilities to overall health and status of C2 systems. In return, EABO must contribute to reliability and resiliency of C2 networks across all domains to ensure completion of mission essential tasks. Reliability, resiliency, and *graceful degradation*[30] based on network and data prioritization are achieved through sound design of the overarching naval C2 architecture, effective integration of the systems that compose that architecture, and effective training of network operators in the human and automated procedures at the tactical level.

Littoral forces must understand how the adversary intends to target and engage friendly C2 systems, and they must actively develop appropriate responses to mitigate the adversary's potential effects on C2 systems. The goal is to have a naval C2 architecture that can absorb adversary effects yet continues to function effectively and support mission objectives.

To perform "Assure Command and Control and Critical Systems", littoral forces conducting EABO must accomplish several key tasks:
- Understand friendly C2 requirements and SOM
- Understand adversary capabilities and TTPs for affecting our C2 systems
- Monitor in near real time the overall health and status of C2 systems and other critical systems
- Monitor in near real time the threats and vulnerabilities to C2 systems and other critical systems
- Provide alerts when critical systems are compromised or become vulnerable to exploitation or attack
- Control and monitor defensive cyberspace operations (DCO) in near real time and employ DCO internal defensive measures (DCO-IDM)
- Monitor and map electromagnetic and cyberspace signatures emanating from the use of C2 systems and other critical systems
- Monitor the use of EMS-dependent systems to identify and minimize EMS fratricide

4.4.2 Provide Information Environment Battlespace Awareness

EABO contributes to naval power projection by augmenting the Navy's fleet sensor network and enhancing understanding of the information environment. Understanding IE vulnerabilities, threats, opportunities, and their potential strategic, operational, and tactical impacts enables flexible responses across the competition continuum. Providing flexible response options in the IE and across all domains requires the ability to gather and fuse relevant information from as many sources as time and resources permit.

Expeditionary advanced base operations will impact the IE in both intentional and unintentional ways. These impacts will affect all domains, influence both local and global audiences, and shape the perspectives of friends, foes, and neutral parties. Understanding these impacts will support the development of potential responses in the IE in support of EABO and will be essential to mission accomplishment and successful naval operations.

Littoral forces must assess IE-specific conditions across all domains. This includes understanding how littoral force physical, technical, and administrative signatures are seen and understood by nearby observers and by the adversary seeking intelligence or a firing solution. This effort requires the fusion of assessments and near-real-time actions to have as complete of a picture as possible of the IE and OE in relation to each other. This effort should result in the planning and execution of deliberate blue-force activities across all domains with an understanding of potential effects in the IE. This requires measuring and assessing blue-force activities to determine IE effects and understand how conditions change.

---

[30] The term "graceful degradation" describes systems that continue to run at some reduced level of performance after one or more components fail. It is a level below "fault-tolerant" systems, which continue operating at normal speed and performance despite a component failure.

Littoral forces must accomplish several key tasks under "Provide Information Environment Battlespace Awareness":

- Develop and maintain a running estimate of the IE
- Incorporate the following types of threat information into the IE running estimate: intelligence, indications, and warnings regarding technical, organizational, or human targets and target system parameters; target C2 networks and nodal dependencies; cyberspace-operations capabilities and actions; space-operations capabilities and actions; and EMS operations (EMSO) capabilities and actions
- Incorporate the following types of environmental information into the IE running estimate: foundational geospatial intelligence; meteorological information; electromagnetic environment; information on the cyberspace physical layer, cyberspace logical layer, and cyberspace persona layer information (to include social media information); open-source intelligence (OSINT), which includes analysis of local, regional, and global media; visual information (VI); enemy satellite surveillance identification; and civil information
- Incorporate the following types of friendly force information into the IE running estimate: health, status, and vulnerabilities of C2 systems and critical systems; electromagnetic and cyberspace signatures emanating from the use of C2 systems and critical systems; communications strategies of adjacent headquarters, higher headquarters, and US government (USG) agencies within the AO to prevent information fratricide; plans, orders, and coordination instructions; instances of EMS fratricide; and the position, location, payload, and mission of assets across the battlespace
- Fuse naval and joint-force information activities with competitor, adversary, enemy, and neutral information activities to identify vulnerabilities, threats, and opportunities in the IE

4.4.3 Attack and Exploit Networks, Systems, and Information

Maritime power projection includes "a broad spectrum of offensive military operations to destroy adversary forces or logistic support or to prevent enemy forces from approaching within enemy weapons' range of friendly forces."[31] EABO supports maritime power projection and other naval functions by targeting and engaging adversary networks, systems, and their underlying information through both traditional combat power and military information power. EABO potentially increase the range and lethality of traditional strike effects relative to standoff capabilities alone, while also providing "avenues of approach" to information-specific targets across all domains and the EMS. Effective attack and exploitation of enemy networks, systems, and information requires a detailed technical target system analysis on the adversary's kill chain. Close and early collaboration with intelligence is critical to understand nonlethal vulnerability characteristics (NVCHAR), and associated nonlethal reference points (NLRP), illuminating ways and means for littoral forces to affect the adversary's use of the EMS.

By their proximity to adversary networks, systems, and information, littoral forces conducting EABO gain access and provide flexibility in targeting and the execution of specific missions. Naval forces target adversary critical information dependencies through both technical and nontechnical means. Technical means target the adversary's data and underlying information networks, while nontechnical means target the human and social elements of the adversary's decision-making system. Planning considerations can include how to interrupt the flow of information or inject information at the right time, place, and duration to gain an operational advantage. By interrupting the flow of information, commanders can affect the quality or timeliness of a decision made by an adversary, prevent a decision from being made, or prevent delivery to those who must act on it.

---

[31] Joint Chiefs of Staff, *DOD Dictionary of Military and Associated Terms* (Washington, DC: US Department of Defense, 2022).

The application of the information capability areas described below in section 4.5 provide potential pathways to attack the adversary's ability to make decisions or take action. Traditional combat power capabilities now include attacking adversary networks, systems, and information, informed by NLRP intelligence information and synchronized with Naval maneuver. Determining how and when to engage targets, however, must be balanced with the EABO requirement to manage signatures and persist forward in contested environments.

Under the "Attack and Exploit Networks, Systems, and Information" task, littoral forces must execute the following:
- Plan, synchronize, and employ multi-domain IE capabilities in support of and synchronized with the SOM in near real time throughout the AO
- Monitor, in coordination with higher headquarters (HHQ), fleet forces, and the joint task force, EW operations, offensive cyberspace operations (OCO), and intelligence operations (including signals intelligence [SIGINT]) to avoid EMS fratricide and recommend alternative COAs or combined-arms solutions to achieve the desired effects
- Identify, nominate, conduct and coordinate information activities that target enemy networks, systems, and information
- Coordinate with HHQ agencies to ensure a shared situational awareness of human dynamics, social network links and nodes, atmospherics, environmental characteristics, and personal intent
- Submit and deconflict requests for support (RFSs), such as electromagnetic attack requests and space support requests, throughout the targeting process
- Deconflict EMSO-related RFSs with the C2 architecture, intelligence collection plan, planned cyberspace fires, airspace control order, and the joint restricted frequencies list (JRFL)

4.4.4 Inform Domestic and International Audiences

Inform operations seek to affect the knowledge, perceptions, attitudes, and behavior of publics to support attainment of friendly force objectives while maintaining institutional reputations. The adversary will seek to discredit US operations by conducting aggressive propaganda operations to swing local, regional, and global public opinion against US naval forces. EABO require dedicated operations to identify and inform various audiences, publics, and stakeholders with clear and accurate communications to build understanding and support for operational and institutional objectives, as well as to counter false narratives. Such inform operations in support of EABO require constant integration with HHQ communications assets and day-to-day engagement with key publics to reinforce strategic messaging and to reassure local, regional, and global partners and allies.

In day-to-day campaigning and throughout the competition continuum, it is a vital concern of littoral forces to understand the impacts of their operations on the local population and environment. Inform operations must also aggressively preempt, counter, and mitigate adversary disinformation and propaganda campaigns aimed at undermining friendly operations. In coordination with HHQ and given appropriate authorities, littoral forces must execute a deliberate and coordinated communications strategy targeted at key public and other relevant audiences to create a sufficiently permissive operating environment to enable employment of combat capabilities at the desired time and place.

Communication strategy and operations (COMMSTRAT) Marines create information products that support all levels of war for release to audiences, publics, and stakeholders to reinforce strategic messaging. Whether through print or digital media, official messaging from littoral force leaders serves as a means to both connect friendly-force activities to a larger strategic purpose and counter adversary propaganda and disinformation. COMMSTRAT also maintains institutional and command reputation by correcting misinformation to preserve credibility with the US population, which may include national-level decision makers.

Key leader engagements (KLEs) provide direct and valuable person-to-person communication. Careful consideration should be given to not only which leaders and audiences to engage but also when to conduct engagement. EABO objectives will help determine the purpose and intent of KLE and allow decision makers to determine how best to engage, whether through direct leader-to-leader dialogue, media engagements, or other opportunities to inform key audiences in support of EABO objectives.

To perform "Inform Domestic and International Audiences", littoral forces conducting EABO must accomplish several key tasks:
- Integrate and synchronize with HHQ COMMSTRAT, joint force public affairs, and regional USG communication efforts (e.g., public diplomacy officers at US embassies to conduct strategic communication and assess communication effects) to develop a command narrative
- Acquire surface, subsurface, and aerial VI to document EABO as required
- Conduct, coordinate, and monitor in near real time EABO actions that will have effects in the IE, which include physical attacks, force movements, psychological operations (PSYOP), and all civil affairs (CA) operations
- Acquire, process, and transmit operational imagery in a command and control denied or degraded environment (C2D2E)
- Create and disseminate visual information/communication products in a C2D2E
- Establish a clear release authority process with HHQ to ensure communication at relevant speeds
- Ensure close integration and synchronization with influence operations to prevent information fratricide

4.4.5 Influence Foreign Target Audiences

Military competition with an opponent is inherently information-centric. Influence operations seek to affect perceptions, attitudes, decision making, and behavior to support attainment of friendly force objectives. For example, naval power projection is a strategy aimed at global deterrence through forward-postured naval forces.[32] Such deterrence is an effort to influence competitors and adversaries to limit actions detrimental to our strategic objectives. EABO support this effort by providing and maintaining the ability to posture naval forces forward. Influence operations, in turn, should be designed to reinforce national strategic objectives and create a permissive environment for EABO.

Littoral forces must operate with the understanding that every action in the OE has the potential to influence target audiences in the IE. The physical maneuver of forces, civil-military operations, and military information support operations (MISO) are key contributions to shaping the EABO environment. During operational design, these activities must be understood for their value in influencing target audiences and incorporated into the EABO planning process.

To perform "Influence Foreign Target Audiences", littoral forces conducting EABO must accomplish several key tasks:
- Monitor MISO
- Coordinate MISO message and theme dissemination
- Coordinate CA operations
- Support ongoing assessments of all influence operations
- Maintain awareness of the target list, nominate effects against approved target audiences in the targeting process, and coordinate influence operations with HHQ's fire-support plan

---

[32] OPNAV, *Naval Warfare*, NDP 1.

- Coordinate operations with tactical psychological operations teams and organic capabilities for the dissemination of audio, visual, and audio-visual messages, including by loudspeaker, leaflet and other print products, face-to-face engagements, and radio broadcasts
- Coordinate operations with expeditionary COMMSTRAT operational support teams (OSTs)
- Coordinate organic capabilities, such as directed imagery capability, for the acquisition, production, and dissemination of both written and visual information, including by loudspeaker, leaflet and other print products, face-to-face engagements, and radio broadcasts
- Coordinate COMMSTRAT and expeditionary COMMSTRAT OST support with intelligence requirements development, intelligence collection, surveillance, reconnaissance, targeting, site exploitation, surveys, and combat assessment
- Perform target audience analysis and recommend and prioritize audiences for development and engagement
- Gain long-term, persistent, and consistent access to relevant audiences
- Plan, coordinate, and execute deliberate presence, profile, and posture activities

4.4.6 Deceive Adversary Decision Makers

Deception actions deliberately mislead adversary decision makers. They aim to cause their targets to take or not to take specific actions, thereby contributing to attainment of friendly force objectives. This task differs from the influence task in four main ways: (1) by the intended effect, (2) by a focus on a much more limited and specific target audience, (3) by a focus on the target's sensing capabilities, and (4) in the authorities required to execute the actions. Succeeding in this task requires the integration of physical actions with specialized capabilities.

Littoral forces must accomplish several key tasks under "Deceive Adversary Decision Makers":
- Coordinate and ensure deception actions are aligned and integrated with higher-level plans
- Plan, coordinate, and execute deception in support of operational security (OPSEC) and tactical deception
- Coordinate the timing and tempo of deception actions in support of tactical-level commanders and ensure deception actions are synchronized with other operations that may affect or be affected by deception actions (e.g., EMSO, OCO, DCO, COMMSTRAT, fires, and maneuver)
- Maintain awareness of efforts by counterintelligence personnel to protect against threats from outside the unit
- Plan, coordinate, and manage the use of administrative, physical, and technical signatures

4.4.7 Control Information Capabilities, Resources, and Activities

This task is vital to creating and leveraging military information power. It is through this task that information maneuver forces capabilities, resources, and activities are synchronized with all operations. The multidomain nature of the IE requires vertical and horizontal coordination in both time and space. Awareness, timing, and close coordination with all other warfighting functions throughout planning and execution are critical.

Information maneuver forces in support of EABO necessitate synchronization among the coordination elements of the littoral force and both higher and adjacent forces to ensure alignment with mission objectives. Proper control and integration of information activities requires littoral forces to host and participate in various coordination and planning cells to support a daily battle rhythm.

Littoral forces must accomplish the following key task under "Control Information Capabilities, Resources, and Activities":

- Track, monitor, and display all information activities relevant to the littoral force mission and operations
- Coordinate and/or control the employment of information maneuver forces per HHQ direction
- Provide near-real-time re-tasking and reprogramming recommendations based on the mission, emergent battlespace conditions, and capabilities and disposition of friendly forces
- Maintain an updated list of the commander's information-related authorities
- Coordinate the timing of information activities in support of naval and joint operations
- Coordinate emergent requirements and requests for reach-back support, including naval, joint, or national support requests

4.5 CREATING AND EXPLOITING INFORMATION ADVANTAGES

The following information activities and related operations can be used to help units create and exploit information advantages and are aligned to either technical or cognitive disciplines. The technical areas include EMS operations, cyberspace operations, and space operations, while the cognitive areas include inform operations, influence operations, and deception operations.

4.5.1 Electromagnetic Spectrum Operations

Military operations in the EMS involve the transmission and reception of electromagnetic energy in the electromagnetic operating environment (EMOE). EMSO are military actions undertaken by a force to exploit, attack, protect, and manage the EMOE. The EMS is a maneuver space and conducting maneuver within it is similar to maneuver in other domains, requiring three-dimensional positioning and time. During cooperation and competition, EMSO are conducted to ensure adequate access to the EMS. As situations escalate towards armed conflict, EMSO shifts to achieving EMS superiority.[33]

EMSO actions to exploit, attack, protect, and manage the EMOE rely on personnel and systems from the EW, EMS management, intelligence, space, and cyberspace mission areas. EABO will be conducted in an EMOE that is constrained, congested, and contested. Throughout the competition continuum, littoral forces will conduct EABO with denied, degraded, or disrupted access to the EMS. It is paramount that EABO incorporate dedicated EMSO planning and execution to incorporate electromagnetic attack in naval-force power projection while understanding risks to mission and force when EMS access is nonpermissive. To mitigate adversary and neutral actors' actions within the EMS, littoral forces must understand EMS maneuver, SIGMAN, and the larger naval campaign.

Successful planning and employment of EMSO must consider several factors:
- TTPs for fusing ISR feeds and EMSO payloads with lethal and nonlethal fires to compress the sensor-shooter "kill chain"
- Authorities available to plan, conduct, and coordinate EMSO
- Agile/dynamic employment of EMSO, including coordination of EW, collection, and communications plans, to identify and minimize potential electromagnetic interference (EMI)
- Coordination of electromagnetic attack requests with C2 architecture, intelligence collections, cyberspace fires, airspace control, and the JRFL
- EMSO timing and tempo coordinated with SIGMAN plans

---

[33] Joint Chiefs of Staff, *Joint Electromagnetic Spectrum Operations*, JP 3-85 (Washington, DC: US Department of Defense, 2020).

4.5.2 Cyberspace Operations

Cyberspace operations include three distinct efforts: Department of Defense information network (DODIN) operations, DCO, and OCO. DODIN operations and DCO are the most common and are critical to assuring C2. Additionally, DODIN operations should provide redundant access to enterprise cross-domain solutions across networks, ensuring forward elements maintain IE and battlespace awareness to maintain initiative and seize emerging opportunities.

Given the proximity of EABO to adversary infrastructure and units, the cyberspace domain presents both opportunities and vulnerabilities. DCO can be conducted to assess, build, and protect local infrastructure in the operational area to deny adversary access. Select OCO can be conducted to prepare the environment for achievement of strategic, operational, and tactical objectives. As with EMSO, however, any OCO must be carefully planned in advance and executed in accordance with SIGMAN principles, allowing the adversary to see or exploit only carefully selected signals and capabilities because they offer possible exposure of further adversary vulnerabilities.

Successful planning and employment of cyberspace operations must consider several factors:
- TTPs and capabilities required to continuously monitor the health and status of C2 networks and critical systems
- Tactical network design that emits minimal unique electromagnetic signatures
- Tailored intelligence to support full-spectrum cyberspace operations and identify key terrain in cyberspace
- TTPs and capabilities required for DCO-IDM efforts to assess friendly networks, weapon systems, and C2 systems based on identified key terrain in cyberspace that are likely targets for exploitation
- TTPs, CONOPS, and authorities required to coordinate and employ OCO assets to defeat enemy capabilities that target C2 and critical systems
- Coordination of cyberspace operations with EMSO and other special capability to minimize friendly EMI
- Coordination of RFSs throughout the targeting process

4.5.3 Space Operations

Space operations are those operations impacting or directly utilizing space-based assets to enhance the capabilities of US forces. DOD space forces are the space and terrestrial systems, equipment, facilities, organizations, and personnel, or the combination thereof, necessary to conduct space operations. Space operations exist to provide space capabilities to the joint force. Space capabilities include space control; positioning, navigation, and timing (PNT); ISR; satellite communications (SATCOM); environmental monitoring; and missile warning. The ability to leverage these space capabilities relies on access via the EMS and cyberspace to systems that provide the desired space capabilities.[34]

Littoral forces must consider space operations early in planning for the conduct of EABO and be proactive end-users of space-based capabilities to effectively shoot, move and communicate. Adversary action in space is inevitable, and the adversary will generate effects that deny, degrade, and disrupt the space operating environment. Understanding when and how the space domain is likely to be affected informs decision makers of the availability of space capabilities and risk to mission and force. Littoral forces must understand how to conduct EABO when certain space capabilities are denied, degraded, or disrupted. Cyberspace and EMS superiority underpin all aspects of space operations; every space-based asset is reliant on the EMS to pass information across the battlespace. Coordination with cyberspace and

---

[34] Joint Chiefs of Staff, *Space Operations*, JP 3-14 (Washington, DC: US Department of Defense, 2018).

EMS planners is required to understand impacts to space due to adversary, friendly, and neutral actions in the EMS and cyberspace.

Successful planning and employment of space operations must account for several factors:
- Understand EABO dependence on space systems and associated vulnerabilities in three satellite system segments[35]: space, control and user
- TTPs and capability to leverage space-based resources to support EABO
- Coordinate EABO requirements for SATCOM and PNT
- TTPs and capability for obtaining space-domain awareness of ISR satellite capabilities to support collections and SIGMAN
- TTPs for proactive use of space-based capabilities to effectively shoot, move, and communicate
- Adversary NVCHAR/NLRP information to inform nonlethal targeting

4.5.4 Inform Operations

Inform operations shape perceptions by correcting misinformation, establishing facts, and putting fleet actions into context. Through the official release of information via traditional media, social media, and face-to-face engagements, these operations correct inaccuracies and discredit adversary propaganda with counter narratives.

Littoral forces, in coordination with fleet and service headquarters, communicate with internal, domestic, HN, coalition, international, and adversary audiences to support strategic, operational, and tactical objectives. These efforts are crucial to enhancing situational awareness and C2. They must be incorporated within the operational planning process to ensure integration in support of EABO objectives. Inform operations are the responsibility of all personnel who make up the littoral force, but personnel specializing in COMMSTRAT, information and knowledge management, and civil affairs will primarily plan, coordinate, and execute the inform operations strategy.

Successful planning and employment of inform operations must consider several factors:
- Knowledge of HHQ narrative, which requires integration with joint force and regional USG communications efforts
- TTPs and capability to gather and analyze publicly available information
- Authorities to microtarget and release messages to the local HN population
- Coordination with influence operations and military deception (MILDEC) operations to prevent information fratricide
- Measurement of communication effects, which requires the use of network and/or application-based data visualization and exploration tools that extract meaning and sentiment from any written content, including news, social media, and online forums to build awareness, spot trends, manage issues, and mitigate risk to operations and institutional reputation
- Status of release authority for public information

---

[35] ***Space Segment:*** It consists of a satellite or a constellation of satellites orbiting the earth and receiving or generating the data and exchanging them with the other segments using appropriate signals such that the targeted applications are fulfilled.
Control Segment: This segment consists of resources and facilities on ground that monitors, controls and maintains the space segment satellites. It sends control and configuration commands to the space segment units to carry out any specific task, reallocate the resource onboard, to maintain or discipline the satellite on its orbit or merely for the management of the data traffic. It also receives the health, housekeeping and other telemetric data from it.
User Segment: It consists of the users who interact with the satellite and exchange data with it through the signals to accomplish the intended applications.

- Ability to create and disseminate information products in an emissions controlled environment
- Support to operational assessments

4.5.5 Influence Operations

Influence operations seek cognitive effects in the human aspects of the IE. The goal is to shape perceptions in the IE to drive behavior change. To plan an effective influence campaign, operations to influence various actors in the environment must account for culture, life experience, social relationships, ideologies, and the influence of those within and outside of the actor's group. Influence operations include, but are not limited to; OPSEC, MISO, SIGMAN, civil affairs, and other special technical capabilities.

All littoral force actions, whether physical movement, maneuver in the OE, or specific actions in the IE, have the capacity to influence key adversary and neutral decision makers to the benefit of fleet operations.

Successful planning and employment of influence operations must account for several factors:
- Knowledge of the HHQ narrative
- TTPs and capability to gather and analyze publicly available information
- Authorities at the local level to release MISO messages to various audiences, including the local population of the HN
- Coordination with public affairs, civil affairs and civil-military operations
- Support to development of the protected target list
- Support to MILDEC, SIGMAN, and OPSEC
- Support to target audience analysis and operational assessments

4.5.6 Deception Operations

Deception operations seek to mislead adversary decision makers, thereby causing the adversary to act or not act in a manner beneficial to friendly force objectives. Success in this area requires the integration of physical actions with specialized capabilities to deliberately mislead the adversary.

Successful planning and employment of deception operations must consider several factors:
- Alignment and integration with HHQ plans
- TTPs, capabilities, and authorities to conduct deception operations
- Coordination with inform and influence operations to prevent information fratricide
- Measurement of deception effects

4.6 INFORMATION MANEUVER FORCES

All personnel involved in executing EABO have a role in the day-to-day generation, preservation, denial and projection of information. There are, however, information maneuver forces whose primary responsibilities are to plan, coordinate, and execute operations within information specific disciplines. These specialists ensure space, cyberspace, EMS, inform, influence, and deception operations are synchronized with the larger operational design and aligned with strategic objectives. Littoral forces must deliberately test and evaluate which information disciplines can and cannot be executed effectively with current force structures. This assessment helps to determine the right division of labor between forward-deployed enablers and specialists providing reach-back support from outside the immediate operating area.

The littoral force's proximity to local and adversary infrastructure is a primary benefit of EABO because it provides unique access and targeting opportunity in support of fleet operations and denial of adversary capabilities. A major challenge for the littoral force is to determine the most effective means of integrating space, cyberspace, EMS, and cognitive effects into the operational planning process. There are several options for this integration. The most satisfactory method is for the primary information coordinator to have equal standing with primary staff officers of the littoral force, thus maintaining balance in the representation of all warfighting functions during planning and execution. An alternative is to fully integrate information maneuver expertise with the littoral force operations section, either as a standalone cell or within the Fires and Effects Coordination Cell. Littoral force commanders should experiment with the staff structure to find the best combinations for integrating information into operational planning.

The unique characteristics of EABO challenge the littoral force to determine the most effective task organization and table of equipment for EABO missions. Operational requirements must drive the appropriate mix of information-specific enablers included in forces conducting EABO. The following enablers should, at a minimum, be considered.

Information Lead Planner. The Information lead planner serves as a focal point of information planning and execution on the littoral force staff. This specialist:
- Integrates the information warfighting function into operational design, planning, and execution;
- Synchronizes information activities;
- Manages day-to-day employment of information maneuver forces in support of EABO.

Multi-Domain Effects Team. A small team of select subject matter experts who are organized, trained, and equipped with emerging technologies designed to leverage littoral forces placement and access.

Cyberspace Mission Elements (CME). Littoral forces determine the most effective location from which cyberspace operators support EABO missions. In many cases, CME may best support the littoral force from a position geographically removed from the littoral force main headquarters. The CME:
- Plans, integrates, and coordinates DCO and DODIN operations in support of tactical and operational units;
- Executes OCO to achieve specific tactical, operational, or strategic objectives when provided the capabilities, responsibilities, and authorities to do so.

Communications Strategy and Operations Team. COMMSTRAT teams include a mix of personnel capable of communication counsel, research and analysis, planning and integration, assessment and evaluation, media engagement, issue management, crisis communication, imagery acquisition, and multimedia product creation and dissemination. Due to their ability to engage in the IE directly, daily, and globally, they can significantly and favorably impact the IE with accurate, truthful information that reinforces institutional credibility. The COMMSTRAT team:
- Integrates with HHQ COMMSTRAT to identify audiences, publics, and stakeholders in the AO
- Provides accurate, truthful, timely, and relevant information to identified public audiences
- Provides capabilities in surface, subsurface, and aerial photography; motion media; graphic design; and reproduction in support of operational and planning requirements

Civil Affairs Team. Civil affairs is the commander's conduit between littoral forces and the local civil authorities. The civil affairs team:
- Conducts area assessments, audience analysis, and early entry civil reconnaissance and civil network development as a part of operational preparation of the environment (OPE)

- Establishes, maintains, influences, and exploits relations among military forces, governmental and nongovernmental organizations (NGOs), and the civilian populace in pursuit of mission objectives[36]
- Facilitates access and maintains permissive relationships in the forward operating environment

Psychological Operations Team. PSYOP teams are capable of supporting influence operations at the tactical, operational, and strategic levels. For the littoral force, the PSYOP team:
- Conducts area assessments, audience analysis, and early entry civil reconnaissance and civil network development as a part of OPE
- Influences adversary and foreign audience behavior in a manner beneficial to friendly force actions and objectives
- Facilitates operational and tactical actions to further littoral force ability to influence foreign audiences
- Supports military deception operations

Electromagnetic Warfare Team. EW missions are executed from a variety of platforms across multiple domains. EW enablers provide necessary experience and expertise to conduct offensive and defensive EW in support of the littoral force. The EW team:
- Plans and executes operations designed to shape, limit, disrupt, exploit, or attack the adversary's access to and use of the EMS
- Collect and analyze data from the EMOE to inform a Commander's decisions
- Protects friendly freedom of action in the EMS

4.7 ALIGNMENT AND INTEGRATION OF INFORMATION IN EABO

Information activities are not planned and do not occur in isolation. Enablers from all information areas shall operate with an understanding of the available support from, and their responsibilities to, the larger information enterprise. Coordination with higher echelons is essential to both (1) ensuring proper authorities to execute specific activities and (2) aligning these activities to operational and strategic objectives.

4.7.1 Higher Echelon Alignment and Coordination

Information activities in support of EABO provide geographic and functional combatant commands the opportunity to leverage forward-deployed information resources and capabilities to directly support strategic objectives. Forces conducting EABO within the adversary's WEZ gain unique placement and access, which may be employed to prepare the environment for future operations. While conducting EABO, littoral forces must exercise and test the connections and relationships with the Navy numbered fleets, the joint force, and the greater information community.

The Marine Corps information warfighting function provides a range of skillsets and capabilities that littoral forces can leverage to conduct EABO. Most information maneuver force specialties and capabilities reside within the MEF Information Group. There are also planning and coordination elements within the service components at the combatant commands. Headquarters, United States Marine Corps provides strategic planning and guidance to the wider Marine Corps information community. Not only can EABO information maneuver forces leverage the collective expertise of the Marine Corps, but lessons learned from operations, activities and investments in support of EABO should drive system and force design and further development of information capabilities across the service.

4.7.2 Naval Integration

The naval character of EABO demands that littoral forces execute actions to gain, maintain and exploit information advantages in close coordination with fleet objectives. Information planners and enablers should operate in close coordination with, and, in many cases, under the cognizance of the Navy information operations warfare commander (IWC) in composite warfare. Information maneuver forces and units of action in the littoral force should understand that EABO support fleet operations. Actions to gain, maintain and exploit information advantages in support of EABO not only enable successful execution but also support fleet operations. Information maneuver forces with the littoral force must understand their supporting relationship to the IWC and their role in supporting fleet objectives.

The Marine Corps information commands and Navy IW relationship requires extensive experimentation, testing, and integrated training to develop mutual understanding among Marine information maneuver forces and Sailors within the Navy's IW communities. With the Navy's IW missions in mind, littoral forces should test and experiment with emerging technical capabilities and information maneuver forces to mature their ability to support EABO. Follow-on experimentation should examine how littoral and IW forces can more specifically support fleet objectives through EABO. United States Coast Guard international engagements and partner capacity building efforts within the OE will also require deliberate integration in the IE. Finally, two variables are critical to ensuring naval integration for information activities: (1) understanding the authorities needed to provide commanders the flexibility to execute all Information function tasks in support of EABO and (2) leveraging the required processes to obtain approval from appropriate authority level.

4.7.3 Special Operations Force Integration

Special operations force integration provides vital means to generate, preserve, deny or project information, especially in cooperation and campaigning. SOF's unique authorities, relationships, and capabilities provide access and placement to conduct these actions across all functions and capability areas to meet commander's intent. Actions to gain, maintain and exploit information advantages often provide the critical means to compete below the threshold of armed conflict. They also enable and set conditions for EABO and littoral force priorities and lines of effort. SOF increasingly employ information to shape the environment to seize and sustain advantage in competition and enable naval expeditionary forces to win in conflict.

4.8 AUTHORITIES

Effective information activities require multidomain actions executed simultaneously across the competition continuum. Commanders must fully understand their authorities (the power to perform some act or take some action), which is often characterized as permission. They must consider not only things that provide affirmative permission to act, but also those things that restrict their ability to act and where approval of those authorities resides. Therefore, authorities provide the "left and right limits" within which one has freedom of action and dictate the echelon at which authorities are retained.[37]

Authorities to perform specific information functions and employ specific capabilities reside at different echelons depending on whether the littoral force is conducting day-to-day steady state operations or is tasked to execute an EABO mission during armed conflict. The Geographic Combatant Commander (GCC) or JFMCC may direct operations to gain information advantage to set conditions for future EABO missions. Once tasked with an EABO mission, the littoral force commander and staff must leverage

---

[37] Joint Chiefs of Staff, "Insights and Best Practices Focus Paper," *Authorities* (Washington, DC: US Department of Defense, 2016).

organic and enterprise information capabilities to execute the information warfighting function to set conditions in the OE to execute the EABO mission.

Mission-specific command relationships define the authority a commander has over assigned or attached forces. Effective command relationships enable the expeditious and effective employment of space, cyberspace, EMS and cognitive capabilities. In many cases, the authorities for employing capabilities across space, cyberspace, and the EMS reside at levels above the littoral force. Some authorities reside at the combatant command level—or are retained at even higher levels—while others are delegated to the operational commander.[38]

The delegation of authorities to lower levels could provide commanders the flexibility to gain and maintain advantages relative to the adversary. The littoral force must demonstrate during training and experimentation the capability to responsibly employ authorities not currently assigned. Attaining objectives is predicated on possessing the right amount of situational awareness in conjunction with the operational control to execute. These elements together provide the commander an understanding of risk to mission, risk to force, and risk to adjacent activities.

Littoral forces must streamline authorities to the greatest degree possible. EABO within an adversary's WEZ place a high priority on responsiveness to orders and permissive executive authorities. The authorities to execute all aspects of the Information function do not currently exist at the littoral force level. Identifying gaps in authority requirements should be a primary concern of littoral forces in EABO testing and experimentation to ensure littoral forces can execute all information activities and tasks.

INTENTIONALLY BLANK

CHAPTER 5

AVIATION OPERATIONS

5.1 GENERAL

Marine Corps aviation delivers lethal, effective, and survivable capability to enable naval and joint campaigning in all domains across the competition continuum. While operating from austere, distributed locations and across extended distances, the aviation combat element (ACE) must be capable of minimal sustainment, fully networked, and entirely interoperable with the joint force and America's allies and partners. Marine Corps aviation provides a cutting-edge advantage to the naval force through the six functions of Marine Corps aviation. The stand-in ACE will equip tomorrow's MAGTF with credible aviation lethality, agility, and information to compete against and deter our nation's adversaries.

While joint and service doctrine for aviation planning and operations remains constant, there are additional considerations necessary to ensure the maximum effectiveness of Marine aviation supporting EABO. Specifically, aviation support to EABO demands a new concept to describe distributed aviation operations. These adaptations are necessary to littoral forces and contribute to the joint campaign.

5.2 PURPOSE AND SCOPE

This chapter discusses the roles, functions, and tasks of Marine Corps aviation as it relates to littoral forces conducting EABO without focusing on specific aircraft, weapons, systems, or C2 architecture used in execution of EABO.

Littoral operations, like many naval operations, are inherently aviation intensive. Protecting forces ashore and at sea, strike warfare, maritime patrol and reconnaissance operations, assault support, air defense and antiair warfare (AAW) have all proven to be aviation intensive in the modern era; conducting such operations in a contested littoral area is bound to be similarly intensive. Consequently, aviation forces tasked with supporting littoral forces conducting EABO and aviation units organic to these forces must be aware of the unique requirements of littoral operations.

5.3 ROLE OF AVIATION IN EXPEDITIONARY ADVANCED BASE OPERATIONS

Aviation operations in EABO are necessary for distributed forces to leverage the virtues of mass without the vulnerabilities of concentration. Massing distributed effects requires a force that is adept at reconnaissance and counter-reconnaissance, digitally interoperable with the joint force, and physically capable of maneuvering with speed and depth across expansive geographic areas. Marine Corps aviation fills these requirements with critical capabilities that digitally integrate aerial and ground sensing with lethal fires and long-range maneuver and sustainment; enabling the SIF to thrive in a multi-domain, contested environment.

Marine aviation must further its capabilities for operating in a distributed littoral environment as an essential element of the littoral force. Aviation in support of EABO has three key characteristics: the persistent distribution of aviation elements across extended distances; operation of distributed aviation elements with minimal sustainment from rear-areas; and networking distributed aviation elements with the littoral force command and control architecture.

The six functions of Marine aviation remain valid: offensive air support, AAW, assault support, air reconnaissance, EW, and control of aircraft and missiles. Marine aviation will provide support to the LFC across the competition and conflict continuum. This support could be sourced from a range of deployed ACE formations – Unit Deployment Program, Dynamic Force Employment, MEU, or other purpose-built models – meant to confuse and complicate adversary planning while simultaneously improving theater security cooperation with allies and partners across the LOA. Scalable in nature, the ACE can increase in presence and persistence to a fully functional ACE executing all functions of Marine aviation in support of the broader LFC objectives.

The role of aviation in EABO is to support the LFC's mission objectives. The LFC must be enabled to command and control aviation forces, understand the requirements of a significantly distributed force, and how to leverage permeability into and out of the area of operations by aircraft in direct or general support. As with other types of operations, these aviation assets of the littoral force are normally organized as an ACE and placed under the command of the senior Navy or Marine Corps aviation officer with the preponderance of the aviation forces and the ability to C2 all the distributed forces. Effectively, this creates an integrated naval aviation task force or task group supporting the broader maritime campaign in the littorals. The ACE supporting the littoral force will likely resemble a unified naval aviation element, which executes aviation functions in support of the littoral force's missions as the SIF. In supporting these missions, the littoral force ACE commander is responsible for accomplishment or coordination of the following tasks:
- Plan aviation operations and use of airspace
- Plan and coordinate the availability of aircraft, crews, ordnance, fuel, facilities, and vessels capable of flight operations
- Task littoral force aviation assets, including drafting the air tasking order (ATO) and air plan
- Direct employment of littoral force aviation assets and coordinate their employment with joint, coalition, and host-nation aviation assets, capabilities, and resources
- Generate the air tactical picture for the COP
- Serve as the strike warfare commander (STWC) when tasked under composite warfare
- Serve as the AMDC when tasked under composite warfare
- Serve as the EXWC when tasked under composite warfare
- Serve as the airspace control authority (ACA) within a LOA when tasked under composite warfare
- Serve as the air resource element coordinator (AREC) when tasked under composite warfare
- Serve as the helicopter element coordinator (HEC) when tasked under composite warfare
- Generate aviation capabilities supporting the CWC, warfare commanders, functional group commanders, and coordinators under composite warfare
- Plan and coordinate base and missile defense

Refer to chapters 3 through 5 and appendix A of *Composite Warfare: Maritime Operations at the Tactical Level of War*, NWP 3-56, for detailed discussion of functions and responsibilities of the various warfare commanders, functional group commanders, and coordinators as they related to aviation operations.

The LFC will be most aware and equipped to support aviation operations within their LOA. The LFC will be supported by the ACE commander for the use and coordination of facilities, agencies, and capabilities located at sea, in the air, and ashore to execute decentralized control of ACE forces, as well as joint aviation resources supporting the LFC. The ACE commander will be prepared to support the LFC with the tasks that include, but are not limited to:
- Executing aviation operations
- Providing AMD to the littoral force and friendly units in the LOA

- Managing and controlling the air domain within the LOA
- Coordinating with joint, coalition, multinational, and HN air-control agencies
- Executing the littoral force's ATO and air plan
- Providing timely and accurate information to subordinate and adjacent commanders, including commanders operating under composite warfare, to support tactical decisions

5.4 AIR DIRECTION, AIR CONTROL, AND AIRSPACE MANAGEMENT

Refer to chapter 4 of *Aviation Operations*, MCWP 3-20, for a discussion of air direction, air control, and airspace management. These methods of aircraft control and management of airspace enable the LFC and the littoral force ACE to ensure centralized command and decentralized control of Navy and Marine Corps aviation assets as the key enabler to aviation success of EABO.

5.5 FUNCTIONS OF AVIATION IN SUPPORT OF EXPEDITIONARY ADVANCED BASE OPERATIONS

Distributed aviation operations in support of EABO, as with other types of operations, are multifunctional and include the doctrinal six functions of Marine aviation. (Refer to *Aviation Operations*, MCWP 3-20, for a complete discussion of the functions of Marine aviation.) EABO will be aviation intensive, with aircraft often operating at long ranges and high endurance. The mere presence of an adversary WEZ does not obviate the requirement for aviation operating in support of fleet objectives in the context of a maritime campaign. Aviation operations in support of EABO will differ from conventional aviation operations in the following ways:

- Unmanned and autonomous aircraft will be more persistent in a littoral campaign and will be within the ACE in support of the LFC, necessitating more detailed airspace coordination
- The pacing adversaries present a qualitatively and quantitatively more dangerous threat to aviation resources than has been experienced in recent operations
- Projected advances in artificial intelligence, machine learning, and sensor technology—and increased use of tactical data links—will speed targeting cycles and decentralize decision making
- Signatures of aircraft, C2 agencies, aviation capable ships, and supporting bases will require constant management and advances in technology to maintain resiliency

As discussed in section 2.6, EABO C2 methodology may be executed under MAGTF C2, composite warfare, or both. *Composite Warfare*, NWP 3-56, lists AMD, strike warfare, maritime airborne control, offensive counter-air (OCA), defensive counter-air (DCA), electromagnetic attack support, and mobility operations (including aerial refueling). Collectively, these mission areas of Navy aviation broadly parallel the six functions of Marine aviation. However, these mission areas alone do not facilitate functional planning of integrated naval aviation operations in support of a maritime campaign. ACE support to a naval task force or task group will require articulation of the flexible and scalable functions of Marine aviation when aligned to the Composite Warfare construct.

Initial inputs and assessments during the drafting of TM EABO demonstrated the requirement to update the functions to account for the unique elements of operations in the littoral environment and in support of EABO. A proposed modernization of the Marine aviation functions is articulated below and depicted in figure 5-1. These remain tied to the current functions of Marine aviation since these ideas remain valid and timeless. The proposed construct adds one new function and elements of 21st century lethality to the current functions to account for the importance of these missions in an EABO environment. These broad concepts for consideration will require assessment and validation from across Headquarters Marine Corps and the FMF to change doctrine and TTPs. The objective is to reduce planning barriers and better articulate considerations between the Navy-Marine Corps team.

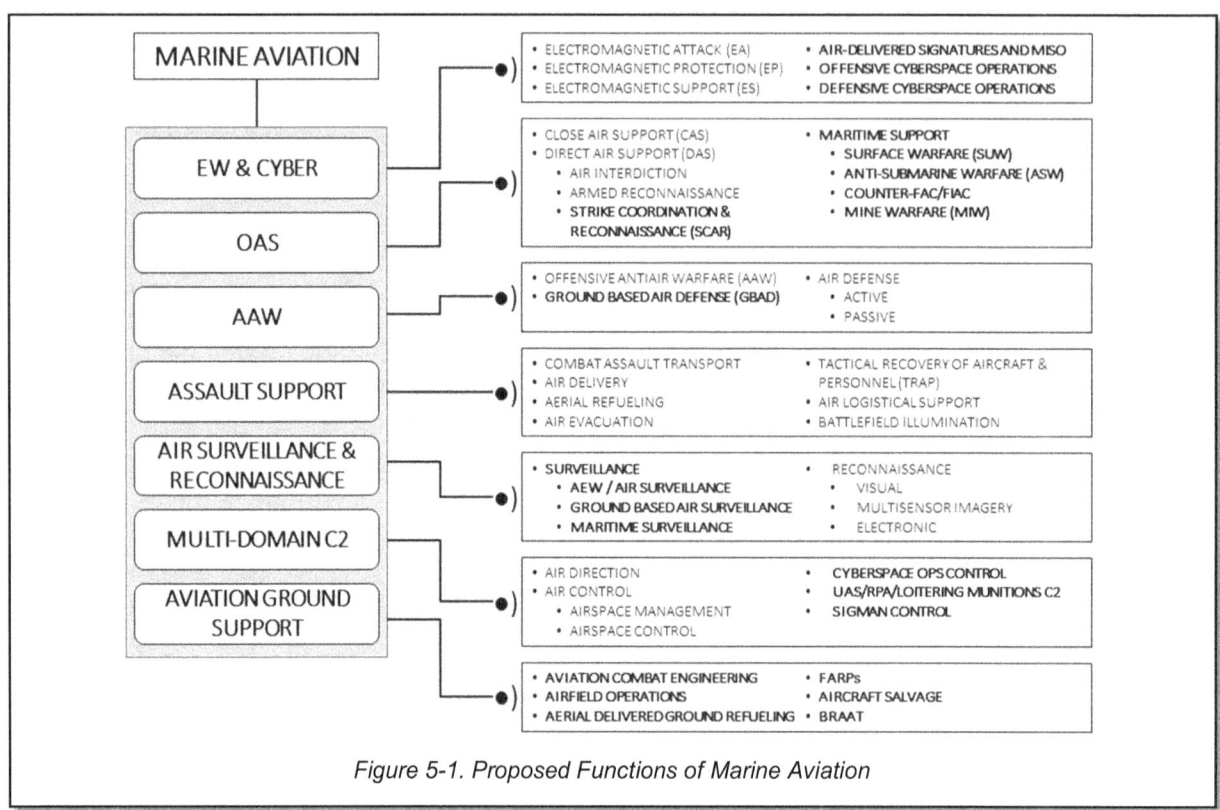

Figure 5-1. Proposed Functions of Marine Aviation

Electromagnetic Warfare (EW) & Cyberspace. This proposed modification expands the traditional EW function to include the full spectrum of cyberspace operations at the tactical through strategic-level. As future advancements of aircraft (manned/unmanned) and the capabilities of their suite of EMS and cyberspace tools modernizes, Marine aviation must be prepared to seek solutions for integrating these areas across the functions of Marine aviation.

Offensive Air Support (OAS). To account for the uniqueness of the naval aviation team fighting alongside each other, this function must be expanded to include strike warfare, surface warfare, antisubmarine warfare, counter fast attack craft/fast inshore attack craft (FAC/FIAC), and mine warfare. The additions of these traditionally Navy missions should initiate a doctrine review for Marine aviation and thus serve as a forcing function to better prepare Marine aviation to support sea control/denial in support of the naval campaign. The adaptation of this function better aligns with terms within the CWC construct and enables more flexible planning and authorities for Marine aviators to better describe critical aviation support to the MAGTF.

Antiair Warfare (AAW). Adding elements of AMD provides added utility for EABO and recognizes this critical element of protection to aviation operations.

Assault Support. This function remains the same.

Aerial Surveillance and Reconnaissance. This proposed function merges aspects of ISR with aerial reconnaissance to provide a future look at the role of Marine aviation, including the future of Unmanned Aircraft Systems functionality to sustain surveillance operations. It also seeks better integration of ground-based sensors used for air surveillance with the objective of a fused situational awareness.

Multi-Domain Command and Control. With new formations and missions comes modified command and control relationships and methods. This function must be updated to account for aviation support to and use of EMS and cyberspace authorities and operations, the addition of proliferated unmanned and potentially autonomous systems, distributed forces in C2D2E comms environment, and control of loitering munitions.

Aviation Ground Support (AGS). The separation of AGS as a function of Marine aviation is critical to clearly outline the roles and responsibilities of logistical and engineer support to an ACE executing EABO. Though non-ACE logistics and engineers will also be enablers to the ACE capacity to leverage capabilities, the resident Marine wing support squadrons (MWSS) and other logistical elements assigned to the ACE will conduct unique tasks to ensure ACE warfighting capacity.

Ultimately, EABO will stress all the functions of Marine aviation. To reduce redundancy across Navy and Marine Corps aviation, a new look at the functions and the authorities necessary for those functions to integrate across the sea services, joint force, and our coalition and allied partners is required. To that end, planners must recognize the high demand for aviation assets, the low-density of Marine assets across a massive theater, and the necessity for flexibility of Marines across the spectrum of Marine aviation to successfully employ all functions of Marine aviation in any environment.

5.6 LITTORAL FORCE AVIATION COMBAT ELEMENT SUPPORTING RELATIONSHIPS

5.6.1 General Support

The support relationship between the ACE and the littoral force is almost always one of general support. This relationship supports the centralized-command and decentralized-control approach to naval operations desired in LOAs. Since demand for aviation support almost always exceeds aviation capacities, the LFC keeps the ACE in general support of the force. This contributes to the most efficient, effective allocation of aviation capabilities, while ensuring effective response to rapidly changing circumstances. The air tasking cycle allocates aircraft to missions and tasks, and is coordinated and passed through C2 architectures. By using the air tasking cycle, the ACE commander allocates finite aviation assets to achieve maximum effect with correct prioritization based on the needs of the LFC.

5.6.2 Direct Support

This support relationship, when set by the LFC, requires the ACE to respond directly to the supported force's requests for assistance. This type of relationship should be established only rarely by the LFC due to the scarcity of aviation assets and the wide range of potential missions that the ACE may undertake in support of the littoral force, especially across vast distances. With the designation of a direct-support relationship, direct communications are required between supporting and supported units, including direct liaison, coordination, and typically local security and logistic support provided by the supported unit.

A littoral force ACE unit in direct support is responsive to the needs of the supported unit. It furnishes continuous support to that unit and its SOM. The direct-support role creates a one-to-one relationship between supporting and supported units. The higher headquarters of the supporting and supported units becomes involved only "by exception." However, each unit must keep its higher headquarters informed of its operations and plans. Examples include an attack squadron in direct support of a subordinate unit of a ground combat element, a helicopter section in direct support of a maneuver battalion, or a low-altitude air defense (LAAD) battery in direct support of a littoral ground unit or maneuver battalion.

5.6.3 Close Support and Mutual Support

Normally, manned and unmanned aviation units do not use close or mutual support. Ground-based and maritime missile and air surveillance units may be placed in a close-support or a mutual-support role relative to other units within the littoral force, or in joint and multinational operations. Additionally, naval personnel conducting AGS and aviation maintenance functions may effectively be in close support while in forward areas.

5.7 LITTORAL FORCE AVIATION COMBAT ELEMENT RELATIONSHIPS WITH THE JOINT FORCE

The JFMCC contributes to and relies on the joint force for the accomplishment of specific functions. Aviation operations, including operations of aviation elements within the littoral force, represent a portion of the LFC's contribution to the JFMCC and the JFC's campaign. Specific littoral force ACE functions in relation to the joint force include the following:

Air and Missile Defense. The JFMCC normally defends the open ocean and littoral regions. This area is allocated to the JFMCC by the Joint Forces Air Component Commander (JFACC). This allocation of air space places the JFMCC air and missile defenses under the area air defense commander (AADC). Under the CWC, the AMDC, as the Alpha Whiskey, fills the regional air defense commander (RADC) or sector air defense commander (SADC) roll, making them responsive to the AADC. In this case, the RADC will coordinate use of aviation assets, C2 agencies, and DCA sorties for the conduct of AMD. Alternatively, the littoral force ACE commander may be designated as a RADC. Such a warfare commander under composite warfare must have facilities and capabilities required for the conduct of AMD. A RADC must also be prepared to employ US Army integrated air and missile defense elements and US Air Force capabilities and sorties for the conduct of AMD in the littorals.

Strike Warfare. Strike warfare includes joint fire support, interdiction, strategic attack, offensive air support, ballistic and cruise missiles, aircraft, littoral forces, and SOF to attack land targets in the operating area. The JFC may task the JFACC or joint force special operations component commander (JFSOCC) to support the JFMCC or the LFC. In this case, the LFC may task the littoral force ACE to assist with air control of JFSOCC or JFACC assets to facilitate strike warfare. Elements within the SIF may employ strike weapons, which must be deconflicted with air defense and air operations by the littoral force ACE commander.

Additionally, a littoral force ACE commander is capable of serving as a JFACC when the LFC is designated as a JTF commander. As such, the ACE commander should act as an integrator of joint aviation for the JTF, with consequent linkages to the theater JFACC and the requirements to bear responsibilities of ACA and air defense commander (ADC), as well as to generate the JTF's air tactical picture and fulfill similar responsibilities outlined in JP 3-30, *Joint Air Operations*.

5.8 LITTORAL AIR COMMAND AND CONTROL

Marine Corps air command and control agencies (MACCs) will be employed by the littoral force ACE commander for the planning and execution of aviation operations and relevant composite warfare operations in support of LFC objectives. Most of these agencies have the common aviation command and control system (CAC2S), consisting of a Link 16-enabled terminal, embedded national tactical receiver, and capable of disseminating and receiving other C2 system COPs / recognized air pictures to the joint force. Commanders must consider the roles, capabilities, functions, and formations of the ACE C2 architecture.

5.8.1 LITTORAL AIR COMMAND AND CONTROL FUNCTIONS

Marine Tactical Air Command Center (TACC). The Marine TACC is the senior Marine air C2 agency, providing the ACE commander with the ability to command, supervise, and direct aviation forces and operations. The TACC is a Marine Corps capability, fashioned out of elements of the Marine air control group, Marine wing headquarters squadron, and augments from across the wing. *Marine Tactical Air Command Center Handbook*, MCRP 3-20F.2, provides detailed information on employing the TACC.

In the context of EABO the TACC should be employed as a scalable and distributed agency. In its legacy employment model, the TACC is not a mobile agency and has a large administrative, electromagnetic, and cyber signature which would limit its survivability inside the WEZ of peer adversaries. When employing the TACC for EABO, planners should consider distributing TACC functions and organizations (current operations, future operations, future plans, air combat intelligence) across the battlespace and pushing command authorities down to the lowest possible level (e.g. launch or divert authority to subordinate MACCS agencies or MAG headquarters). Consideration should also be given to locating some TACC functions outside of the WEZ. When the ACE is employed as part of naval expeditionary force the TACC should be augmented with Navy personnel.

The TACC C2 capability is facilitated by systems such as Composite Tracking Network, CAC2S, and TBMCS, and the Marine wing communications squadron's ability to maneuver the network and assure C2 for the ACE commander. These C2 capabilities ultimately allow the LFC to C2 three-dimensional battlespace, integrate into naval and joint kill webs, and provide a linkage to the LFC command element, JFACC, and JFMCC. Experimentation with alternative TACC employment is required and should be included with fleet experiments.

Navy Tactical Air Control Center (TACC), or TACC Afloat. The Navy TACC is located aboard amphibious warfare ships and manned by Sailors from a tactical air-control squadron. The TACC Afloat provides air direction and air control in the vicinity of amphibious warfare ships, including control of itinerant aircraft, air-support control, and air-defense control. Frequently, Marine Corps aviators and air C2 Marines serve as liaisons in the TACC Afloat. The TACC Afloat is a positive-control agency and can be an effective location from which to control air operations.

Marine Tactical Air Direction Center (TADC). The TADC is a task-organized agency to provide many of the Marine TACC's tasks, but it is employed in a subordinate role to a senior air C2 agency. In this capacity, the TADC may serve in a subordinate role to a Marine or Navy TACC. Due to the task-organized nature and generally smaller signature, a TADC may be a more optimal air command and control agency ashore to the Marine TACC. Detailed information on the TADC is found in *Marine Tactical Air Command Center Handbook*, MCRP 3-20F.2.

Like the Marine TACC, the TADC should closely manage its signatures. TADCs, operating in conjunction with a TACC (afloat or ashore), can provide options for managing signatures and distributing command capacity across the adversary's WEZ, thereby complicating targeting.

Direct Air Support Center (DASC). The DASC is a Marine air-control agency principally responsible for the direction of air operations directly supporting ground forces. The DASC manages several extensions, including the tactical air-control party; tactical air coordinator (airborne), or TAC (A); forward air controller (airborne), or FAC (A); helicopter support team; air support element; and air support liaison team. Due to its lack of organic sensors, the DASC is a procedural control agency. However, if operating with Link 16 capabilities resident in CAC2S, or a joint range extension, the DASC is capable of viewing an air picture provided by other radar-capable agencies and thereby controlling aviation more like a positive control agency. In amphibious operations, it is frequently the first air-control agency ashore.

Direct Air Support Center Handbook, MCRP 3-20F.5, provides detailed information on the DASC and its extensions.

Tactical Air Operations Center (TAOC) and the Early Warning/Control (EW/C) Center. The TAOC and EW/C conduct joint and coalition airspace control and airspace management. These agencies conduct real-time surveillance of assigned airspace; detect, classify, identify, and control the interception of hostile aircraft and missiles; and direct, control, and provide navigational assistance to friendly aircraft. The TAOC and EW/C, when augmented with equipment and personnel, can provide a facility from which a SADC, RADC, or AMDC may operate. The TAOC and EW/C also contribute to the air tactical picture and provide both tracks and radar plots to other missile units and air-control agencies. The TAOC and EW/C use radars (and, in the future, likely passive sensors) to surveil assigned airspace and to generate an air picture. Additionally, the TAOC's and EW/C's radars may be used to contribute radar plots directly to platforms capable of cooperative engagement via the Composite Tracking Network, including Aegis-equipped cruisers and destroyers and E-2D Advanced Hawkeye aircraft. This can extend radar coverage farther ashore than is normally possible using only ship-based radar systems. *Tactical Air Operations Center Handbook*, MCRP 3.20F.6 contains additional information on Marine SADC, TAOC, and EW/C employment.

The TAOC and EW/C present large administrative and cyberspace signatures, and very large electromagnetic signatures (due to the use of radars). In the future, the TAOC and EW/C must make use of passive sensors and emission control (EMCON) measures to the maximum extent possible and must be organized as a tactically mobile agencies to complicate targeting. Active radars must be closely managed to address signature concerns. New tactics must be developed to effectively integrate sea-based radars, ashore radars, and passive sensors to produce a single air picture. The facilities from which TAOC and EW/C crews operate must become disaggregated from the sensors, more mobile, and be capable of operating effectively with fewer personnel than often currently practiced.

Multi-Function Air Operations Center (MAOC). The MAOC is the tactical C2 agency which may be employed in the littorals by the Marine air control group commanders as a multi-function C2 agency. The MAOC combines the air defense, air surveillance and air support functions of the TAOC and DASC into a single, scalable C2 agency capable of employing additional air C2 nodes and teams. These extensions may be either multi-functional or tailored to support land, air, or maritime operations serving as either an extension of the MAOC or operating independently. The MAOC is not a "hybrid" node, rather it takes advantage of pure multi-functionality while optimizing the traditional roles of DASC and TAOC roles in air direction and air control.

The MAOC's mission is to generate an integrated tactical picture to control aircraft and missiles, enable decision superiority, gain and maintain custody of adversary targets, hold adversary targets at risk, and engage multi domain targets as directed ISO Marine Corps, Naval and Joint Forces. Additionally, every MAOC is an aviation command enabler for either the Marine aircraft wing (MAW) or Marine aircraft group capable of performing the current operations functions of the TACC thereby increasing the ACE Commander's options for distribution of TACC functions.

The MAOC is equipped with the air C2 family of systems such as the TPS-80 G/ATOR and CAC2S. The MAOC also employs advanced tactical data systems for joint coordination and fires integration. These systems are integrated on wideband and narrowband tactical transport provided by the Marine wing communication squadron (MWCS) to enable spectrum diversity and increase EMS survivability. Experimentation has shown that the MAOC is even more capable when augmented with TS/SCI systems allowing for intelligence fusion and multi-domain operations.

5.8.2 LITTORAL AIR COMMAND AND CONTROL FORMATIONS

Marine Air Traffic Control Company (MATC CO). MATC CO is the primary Marine air traffic control (ATC) agency capable of providing ATC capabilities at airfields, air sites, and air points. MATC CO contributes to the overall air-surveillance effort, and, in coordination with LAAD, may operate a base defense zone in the vicinity of an airfield. Additionally, MATC CO can coordinate with HN ATC agencies to ensure the efficient and safe integration of naval aviation with HN military and civil aviation in and around air bases. *Marine Air Traffic Control Detachment Handbook*, MCRP 3-20F.7, provides information relating to MATC Co capabilities and limitations.

In the context of EABO, MATC CO(s) should anticipate increased requirements to control austere landing zones, FARPs, and temporary or permanent HN airfields, as well as to conduct ATC liaison tasks.

Low-Altitude Air Defense (LAAD). LAAD battalions are being fielded new equipment sets that expand the threats they can defend against and provide the ability to complete the 'detect, track, identify, and engage' sequence organically. The Marine air defense integrated system (MADIS) provides active and passive detection capability coupled with kinetic and non-kinetic defeat of groups 1-3 UAS, fixed-wing and rotary-wing aircraft. Additionally, MADIS can send and receive air tracks via Link-16 or joint range extension protocols, bringing it into the Joint architecture. LAAD units task organize to meet assigned missions with the lowest employable unit. LAAD Battalions will also be fielded with medium range intercept capabilities which will provide point defense against salvos of subsonic and supersonic maneuvering cruise missiles. The combination of these capabilities contributes to the SIF counter recon efforts during competition and allow EABs to persist inside the enemy's WEZ during times of conflict. *Low Altitude Air Defense Handbook*, MCRP 3.20F.8, provides additional information.

Littoral Anti-air Battalion (LAAB). The LAAB is an element of the MACCS organic to the MLR. It contains aviation C2 and LAAD personnel and equipment and provides the MLR with the ability to conduct AMD, AAW, and air support control in support of the LFC. As an element of the MACCS, LAAB elements will operate in close coordination with ACE air C2 agencies to operate with appropriate authorities and to be able to control and direct air operations.

Other Navy Platforms. US Navy aircraft carriers, cruisers, destroyers, and E-2D aircraft can serve as platforms from which air control and air direction can be performed.

5.9 AVIATION GROUND SUPPORT

Marine aviation has unique logistical and engineer support requirements that enable sortie generation. Support for expeditionary aviation necessitates consideration of these capabilities outside the conventional GCE/LCE support system. AGS enables ACE employment in an expeditionary manner. The MWSS is responsible for providing AGS and does this through execution of the six activities of AGS: forward aviation combat engineering operations, airfield operations, base recovery after attack (BRAAT) operations, airfield damage repair (ADR) operations, FARP missions, and aircraft salvage and recovery (ACSR) operations.

To counter peer and near-peer competitors the Fleet Marine Force must persist and win within the WEZ while operating from dispersed and disaggregated locations. The MWSS must be able to rapidly deploy capabilities from the sea and air to subsequently employ, integrate, and displace while simultaneously generating aviation sorties as part of an integrated naval force. EABO success requires integration and employment of AGS capabilities to support Marine, Naval, Combined, and Joint aviation forces across the competition continuum.

AGS focuses on establishing, maintaining, and repairing expeditionary airfields, landing strips, landing zones, and FARPs. Support can be tailored towards fixed wing, rotary wing, tilt-rotor, and unmanned aircraft. Specialized aviation planning and design is required to accomplish these tasks and is provided by subject matter experts resident within the expeditionary airfield company of the MWSS.

Once an airfield is established, the primary tasks of AGS is providing airfield services to include expeditionary airfield (EAF) services, expeditionary firefighting and rescue (EFR), aviation fuels distribution, and explosive ordnance disposal. The MWSS provides the technical expertise, equipment, and personnel necessary to operate the flight line (e.g., emergency response, aircraft arrestment, aviation refueling, EOD response, managing flight line hours, lighting and marking, and establishing parking).

Another dedicated mission conducted by the MWSS is BRAAT. This is the assessment and restoration of essential airfield operations following an enemy attack involving damage or destruction to the airfield. Aviation units must be restored to the minimum level of combat effectiveness. The objective of BRAAT is to determine the minimum operating strip, which is the minimum amount of area required to launch and recover aircraft. ADR is conducted concurrently with BRAAT, once areas are cleared to begin repair operations. It is initiated to restore an airfield to the minimum operating capability by using materials, procedures, and techniques for rapid repair of damaged operating surfaces to provide for tactical aircraft launch and recovery operations. ADR involves extensive engineer, airfield operations, and coordinated support efforts. Specialized ADR planning is required to ensure the proper personnel, equipment, and materials are available to rapidly restore the airfield to a state of sortie generation. The MWSS is responsible for calculating estimates for repair time, material requirements, and executing the mission.

Specific missions performed by the MWSS include FARP and ACSR. A FARP provides fuel and ordnance necessary for highly mobile and versatile helicopter, tiltrotor, and fixed wing operations. The size of the FARP varies with the mission and the number of aircraft to be serviced. The ultimate objective of a FARP is to minimize response time and decrease turn-around time in support of sustained operations. This is achieved by minimizing flight time to-and-from the refueling and rearming point and reducing the refueling and rearming time. The MWSS is often augmented with personnel from the Marine aviation logistics squadrons for aviation ordnance operations, and Marine aircraft control group personnel to provide air traffic control and communications. Salvaging or recovering an aircraft involves the action of removing an aircraft from a mishap site to facilitate clearance of landing zones, recovery of assets, and repairs to the aircraft. Execution is for the specific purpose of the safe salvage and/or recovery of aircraft without unnecessary damage to the aircraft. The composition of an ACSR mission may vary and each mission requires a planning process where the quantity, minimum operating strip, and billet of each member taking part in the mission shall be determined to meet mission requirements to support the mission.

AGS functions that must receive less emphasis in the future are food service support, aspects of general engineering support, and intrabase motor transport support. It is expected that many of these types of support will be procured through partnerships with host nations and contingency contracting.

5.10 AVIATION PLANNING

Whether operating under MAGTF C2 or composite warfare, the fundamental methods of Marine Corps aviation planning remain valid. For detailed discussions of how to perform aviation planning, refer to *Aviation Operations*, MCWP 3-20; *MAGTF Aviation Planning*, MCTP 5-10A; and *Composite Warfare: Maritime Operations at the Tactical Level of War*, NWP 3-56. Refer to *Marine Corps Planning Process*, MCWP 5-10, for a more general treatment of Marine Corps planning processes.

The overall objective of planning for aviation operations is to reach the optimum balance of efficiency, effectiveness, and flexibility in allocating scarce aviation assets when demand for those assets exceeds supply. In the context of EABO, these planning efforts should strive for aviation operations that demonstrate resilience. The result of such efforts are an air plan, ATO, and supporting aviation documents (e.g., airspace control order, air defense plans, and special instructions).

Broadly speaking, when a littoral force ACE is operating in general support of an LFC:
- The LFC, or CWC when designated, fulfills the role of the MAGTF commander in aviation planning
- The littoral force ACE commander similarly fulfills the role of the ACE commander in traditional MAGTF operations
- The responsibilities of the AREC and the HEC, should the littoral force ACE commander assume this role, are generally performed by their future plans and future operations sections, in coordination with the ATO development cell

5.10.1 Littoral Force Aviation Combat Element in Support of Joint Operations

Per *Doctrine for the Armed Forces of the United States*, JP 1, excess tactical air sorties shall be made available to the JFC, coordinated by the JFACC, to accomplish JFC-desired objectives and effects. Marine Corps forces also have a responsibility to provide three different types of sorties in addition to "excess sorties":
- Long-range interdiction
- Long-range reconnaissance
- Air defense (i.e., DCA)

Given that these sorties were traditionally provided in the context of a land campaign, with the Marine Corps fighting under a service-component headquarters (as in Operation DESERT STORM), or under a joint force land component commander (JFLCC) (as in Operation IRAQI FREEDOM), this arrangement of providing long-range interdiction, reconnaissance, and air-defense sorties may merit renegotiation. Such renegotiation may be desired because during EABO the littoral force ACE is integral to the operations of the JFMCC, and consequently is subsumed by the broader naval aviation forces.

5.10.2 Littoral Force Aviation Combat Element Liaison with Joint/Combined Entities

Liaison between the littoral force ACE and the joint air operations center (JAOC) / combined air operations center (CAOC) will be a requirement. The Marine liaison should be prepared to provide representation to all departments of the JAOC/CAOC, communicate and coordinate between the ACE and joint/combined entities, and provide technical and subject matter expertise concerning how the JFACC, AADC, and theater ACA can best work with littoral force aviation.

In the past, the Navy and Marine Corps each provided separate liaison elements (naval and amphibious liaison element [NALE] and Marine liaison element [MARLE] respectively.) Because the littoral force ACE is an integrated naval aviation formation, operating under the littoral force in support of the JFMCC, the MARLE should be subsumed by the NALE, and the NALE should become a Navy-Marine Corps integrated liaison element. This integrated liaison element should be led by a general officer or flag officer from the Marine Corps or Navy to ensure appropriate weight is carried in joint targeting boards, JFACC apportionment decisions, and coordination and cooperation with the AADC.

INTENTIONALLY BLANK

CHAPTER 6

LOGISTICS OPERATIONS

6.1 GENERAL

"Logistics provides the resources of combat power, positions those resources on the battlefield, and sustains them throughout the execution of operations."[39] Littoral forces rely on resilient and agile logistics that adapt to changing environments and conditions. Persistence, a key characteristic of EABO, is facilitated by incorporating a framework of naval integration, joint logistics enterprise (JLEnt), and Allied and partnered logistics (e.g., coalition; American, British, Canadian, Australian, and New Zealand ; HN; etc....) supporting the movement and sustainment of decentralized forces throughout the littorals. This chapter provides an overview of emerging logistical topics and issues relating to supporting EABO.

6.2 LOGISTICS IN THE COMPETITION CONTINUUM

The Marine Corps operates in competition every day, with the necessity to escalate in response to crisis or conflict at any time. The context of the operating environment has a dramatic impact on the logistics system within an operating area. Planners must understand the impact on logistics across the competition spectrum. The impacts include changes to the availability of capabilities, as well as the orientation of logistics operations and activities across the theater, and potentially across the globe. Logisticians must capitalize on the art and science of logistics to support the national command authority and GCC prior to armed conflict in an effort to prevent lethal engagement.

6.3 TACTICAL-LEVEL LOGISTICS

The JFMCC must be capable of campaigning within competition and transitioning seamlessly to conflict, with the desire to drive conditions back to competition. This exacts a unique demand on FMF units operating outside of the support of a MEF. Marines understand tactical logistics and force posture conducted in support of the JFMCC serve as shaping actions. They assist with development of alliances and expand the JLEnt opportunities. Both of these positive attributes are required if actions transition into conflict but can also serve as a bulwark against this transition. "The ability to sustain lighter, faster, and more distributed operations is critical to maintaining the U.S. naval force's

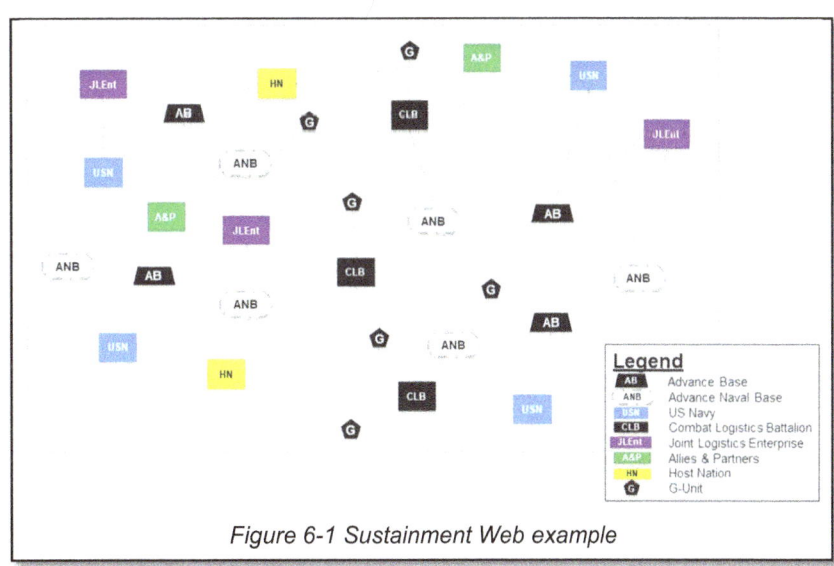

Figure 6-1 Sustainment Web example

---

[39] Headquarters, US Marine Corps, *Logistics* MCDP-4 (Washington, DC: US Marine Corps, 2018).

competitive advantage[40]. The littoral force's concept of logistics support (COLS) should utilize a network of sustainment webs (figure 6-1) that offers diversified distribution, redundant sourcing, and resilient infrastructure via enhanced logistics C2 in an effort to increase JFMCC's COAs. Marines operating within the littoral operating area view tactical logistics as not just a means to self-support but as a strategic and operational maneuver element.

6.3.1 Supply

Minimizing the traditional "iron mountain" requires developing a web of supply sources that are forward, persistent, and capable of meeting the demand of Naval units ashore and afloat at the point of need. This ultimately limits the JFMCC footprint ashore. A balance of organic supply, forward provisioning techniques, access to materiel globally positioned ashore and afloat in the Marine Corps global positioning network (GPN) and Maritime Prepositioned Force (MPF), anticipatory delivery (e.g. "push" logistics) based on data-driven predictive analysis, and operational contract support (OCS) provide commander's options during EABO to reduce their footprint ashore, decrease customer wait time, reduce physical and administrative signature, and increase flexibility in their concepts of logistics support. (figure 6-2)

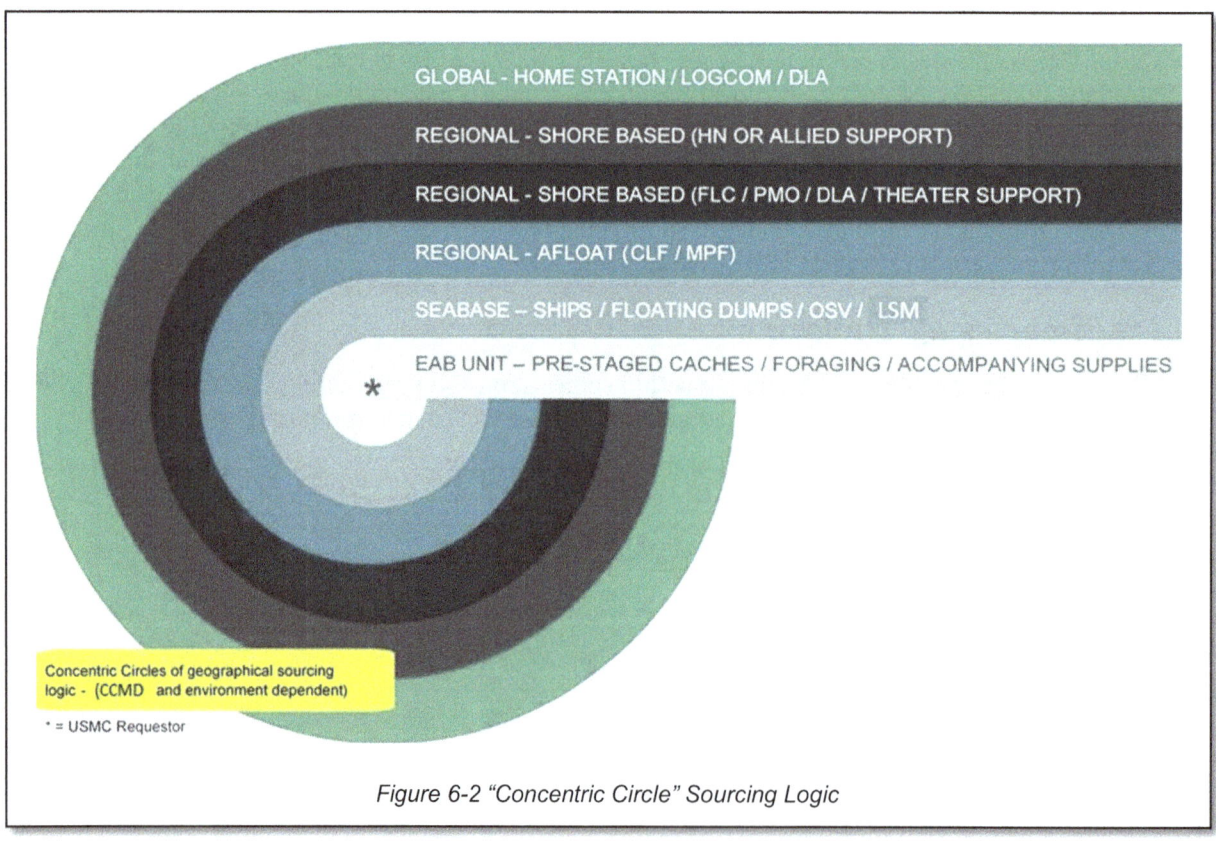

Figure 6-2 "Concentric Circle" Sourcing Logic

The GPN will modernize legacy prepositioning programs into an integrated network of afloat and ashore capability enabling day-to-day campaigning, rapid response to crisis and contingencies, and deterrence, to include support of stand-in forces conducting EABO. Used on a routine basis by forces as they campaign during competition, the GPN enables constant forward presence of elements of the stand-in force as they

---

[40]Irion A., Ekman E., Ryan M., (November 2020). *U.S. Marine Corps Logistics is a shaping factor.* Proceedings, U.S. Naval Institute.

campaign and accelerates force closure during conflict by reducing the deploying forces' fly-in echelon requirements while minimizing demand for strategic and intra-theater lift. Beyond force closure, the GPN will enable forces conducting EABO to persist across the competition continuum by providing sustainment support via globally positioned supplies until theater distribution networks open and materiel begins to flow via the JLEnt.

Commanders should develop sourcing logics for each class of supply, for each specific mission based on the SOM. Time-space analyses for each source is critical during planning as a guide for logistics forecasting, anticipating logistics shortfalls, prioritizing logistics support by node or unit or class of supply, and anticipating the culminating point by node, unit, or class of supply. To achieve flexibility, logisticians and supply personnel must plan and develop resilient logistics supply webs that offer redundancy. It is important to consider "green cell"[41] input of a HN government and civilian population,[42] when developing a COLS that incorporates a multitude of local sources of supply to avoid negatively impacting the economy.

Planners must consider the spectrum of forward provisioning techniques and options available for units during EABO in any theater of operations and at any point along the competition continuum. For example, field ordering officers and unit paying agents at the tactical and micro-tactical levels provide a responsive capability to meet emerging needs quickly and directly through micro-purchases of goods and services. This type of forward provisioning technique will serve as an augment to traditional naval husbandry processes or serve as a replacement during campaigning. The JLEnt offers alternate sources of supply that will reduce demand on the naval sustainment network and limit our footprint ashore. Local sources may also offer an opportunity to reduce signature. The stand-in force offers commanders insight into local economies and patterns of life at the micro-tactical level that will guide sourcing logic development. Supplemental sustainment methodologies included in figure 6-3 represent the flexibility that Marine Corps units must build into concepts of logistics support.

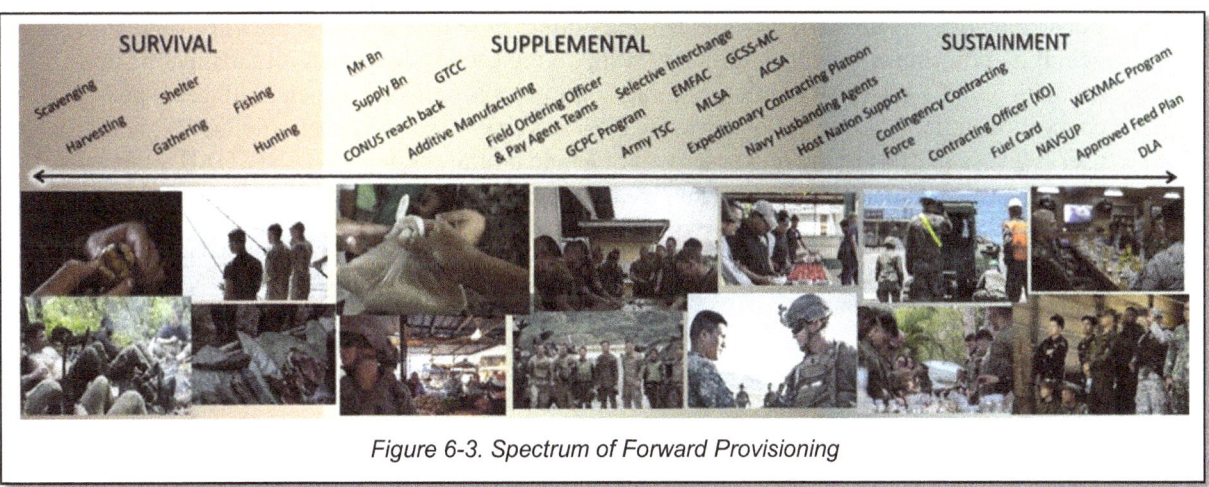

Figure 6-3. Spectrum of Forward Provisioning

Expeditionary Contracting Platoons are home to the Contingency Contracting Force (CCF), task-organized to provide unique capabilities to achieve flexibility, sustainability, and economy. They must consider customer wait time, administrative signature, technical aspects of writing and awarding contracts, and the strategic location of Contracting Officers and technical SMEs. Options like Field Ordering Officer / Pay Agent teams, Government Travel Charge Card holders, Government-wide

---

[41] United States Marine Corps, (June 2020), *Pamphlet 2-0.1, Red Cell - Green Cell.* Marine Air-Ground Task Force Staff Training Program, MSTP Center (C 467), Quantico, VA.
[42] Barak A. Salmoni and Paula Holmes-Eber, (2011) *Operational Culture for the Warfighter: Principles and Applications (Second Edition)*, Marine Corps University Press, Quantico, VA.

Commercial Purchase Card (GCPC) holders, and Acquisitioning and Cross-Servicing Agreements (ACSA) reduce customer wait time, increase flexibility by expanding the supply web, contribute to an integrated deception plan, aid logistics as a form of maneuver, and ultimately reduce the burden on naval logistics chains. Much of this initial work to set the theater should be accomplished by the SIF during routine named exercises and Theater Security Cooperation (TSC) engagements, then captured within easy-to-use logistics systems that enable JLEnt awareness of all available options to use for sustainment.

Purification of available water and local procurement of fuel (made suitable for military equipment with additives), coupled with resilient means of distribution and storage, can reduce the burden on the sustainment network and satisfy demand at or close to the point of need. Additive manufacturing is one method of improving supply responsiveness, reducing demand, and creating efficiency during EABO. To further reduce our footprint ashore, additive manufacturing materials should be locally purchased to the greatest extend possible.

Supply considerations during EABO should include:
- Holistic review of existing joint logistics enablers and support agreements in theater to include ACSAs, Worldwide Expeditionary Multiple Award Contract contractors, Defense Logistics Agency (DLA) hubs, GPN, theater sustainment command locations and capabilities, husbanding agents, approved food sources, approved fuel card and GCPC vendors
- Supply-web and sourcing logic development for predictive analysis
- Distribution platforms by location, theater, and LOA
- Prioritizing forward provisioning techniques within sourcing logics to develop an overarching COLS
- Minimization of storage location footprints, unless incorporated into HHQ deception plan
- Develop creative and uncomfortable ways to reduce consumption rates and reduce demand by challenging doctrinal assumptions based on the SOM and anticipated OE
- Cache network development to include site options; COP development; network standard operating procedures for consuming and resupplying cache sites; cache site ownership and accountability; site security options; storage options based on the OE; camouflaging and concealment; deception sites; risk-worthy sites and risk-worthy contents; SIGMAN; common user logistics platforms and technology; unmanned aircraft systems (UAS) / UAS integration; relocation requirements; and cache site weaponization and self-destruction options
- Administrative signature emitted by supply and fiscal transactions
- Source of supply vetting for expeditionary and contingency contracting opportunities

6.3.2 Maintenance

The littoral force's ability to persist requires positioning of required maintenance capabilities as close to the point of need as feasible or removal of the exquisite maintenance processes through effective supply procedures. The distributed nature of the myriad small, independent maneuver units requires a fundamental shift in approach to maintenance support. To the greatest extent possible, these maneuver units must be equipped with simple, plug-and-play maintainable systems that can be both operated and maintained by the user. The use of virtual reality connecting a school-trained maintainer to a distant user to implement some form of corrective maintenance is one example of an innovative approach. The notion that small combat service support (CSS) detachments will be accompanying these maneuver units to sustain readiness levels of their equipment needs to be discarded.

Although stand-in forces may be able to evacuate equipment via multimodal means to higher maintenance activities for repair, the time and distance required for evacuation reduces the responsiveness of the maintenance system and risks reducing littoral force capability. If equipment cannot be repaired forward in an expeditious manner, then it should be cannibalized, evacuated, or abandoned. Given the anticipated OE, these circumstances are more acute during EABO than many other types of operations.

Materiel in the GPN will include repair parts anticipated to be vital to sustaining operations of forces conducting EABO until access to the JLEnt is established. DLA will continue to provide unique, high demand-low density repair parts and components that local sources of supply cannot replicate or provide due to quality assurance challenges, in accordance with GCC prioritization. The GCC will prioritize DOD joint program of record equipment before Marine Corps peculiar end items. This makes logical sense as it reduces logistics considerations on the CCDR to fulfill multiservice requisitions. This is in direct conflict with equipment required to execute EABO which is more technologically advanced, potentially creating increased demand for civilian field service representatives (FSRs). The ability of FSRs to perform frequent or extensive maintenance as part of the stand-in force presents challenges. Planners must consider increased sustainment and force protection requirements related to an increased role of FSRs supporting stand-in forces, or the USMC must consider the requirement for uniformed personnel to acquire the knowledge, skills, attitudes, and specialized tools that would be required to replicate FSR-capabilities. Consolidating critical skills via multifunctional maintenance training and education will reduce the demand for FSRs forward in the OE. Ideally, forces executing EABO will use commercially adapted equipment that has a global maintenance servicing network and allows littoral forces to access local maintenance facilities and skilled labor. Use of commercial equipment must be scrutinized, not to violate any international laws, should the competition continuum escalate into crisis and conflict.

Finally, Marine Corps commanders must look for ways to support other naval forces and respond rapidly to the JFMCC requirements to repair and rearm navy ships in support of the GCC. It is a very real possibility the most important tactical maintenance supported by Marine Corps commands is not for organic assets but on a destroyer, amphibious warfare ship, Fleet Surgical Team medical equipment, or other littoral force asset in need of an additive manufacture part. Establishing or maintaining EABs with a capability to support other naval forces, as a stand in force, will provide the greatest support to the GCC and JFMCC.

6.3.3 Transportation

The ability of the littoral force to execute effective transportation operations in a distributed environment requires an integrated approach to transportation via sea, land, and air, as well as utilization of both manned and unmanned systems. The transportation of on-call supplies in support of EABO via surface, ground, and/or air platforms provides the MAGTF commander with a multi-modal resupply options. An organic ancillary surface connector with stand-in characteristics of long range, high speed and shallow draft that functions as a heavy truck in the surface domain is one example. The use of uncrewed aerial delivery capabilities to reach remote maneuver units that may be operating in areas void of traditional infrastructure to provide delivery or retrograde of critical supplies may be necessary particularly to link those units to the sustainment aboard littoral surface connectors. Another example is the Joint Precision Airdrop System that provides rapid, precise, high-altitude delivery capabilities when the risk to manned ground, surface, and conventional assault support or helicopter support options are not acceptable or feasible. The littoral force's ability to utilize all modes of transportation and the entire geography within the LOA, for both embarkation and debarkation, reduces the adversary's ability to conduct pattern assessments of the littoral force's distribution network. Coordination across the Naval and Joint Logistics Enterprises is critical to ensure the realization of potential economies of scale, thus increasing the efficiency of distribution networks across one or more LOAs.

Additionally, during campaigning, placement of Marine Corps capabilities, personnel and equipment, will have operational and strategic objectives, even if only for a moment in time. Transportation can be used as an information related capability. Marines operating in the littoral force need only contract transportation for the adversary to respond. Similar to how a minefield or deception minefield can affect decision making, transportation or deception transportation can have a similar impact. For example, how many submarines must a country have to be effective? The answer may be only one because a submarine is everywhere and nowhere at the same time. Transportation can have the same effect.

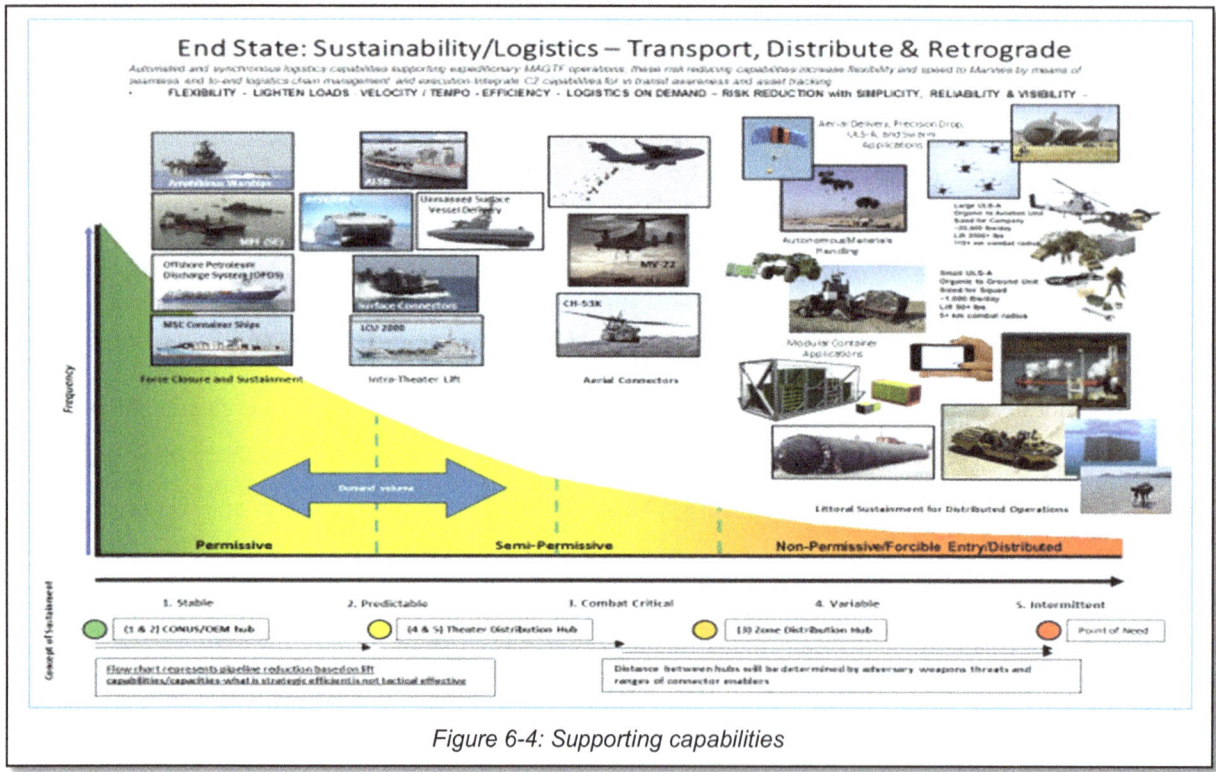

Figure 6-4: Supporting capabilities

Figure 6-4 highlights some of the automated and synchronous logistics capabilities supporting EABO. These risk reducing capabilities increase flexibility and speed to Marines by means of seamless end-to-end logistics chain/web management and execution as well as integrated C2 capabilities for in-transit awareness and asset tracking.

6.3.4 Engineer Operations

Engineering operations are critical to EABO throughout the spectrum of conflict. During campaigning, the littoral force may opt to support allies and partners with local infrastructure improvements, which can mask construction of dual- use infrastructure that will enable conduct of future distributed operations as well as contribute to deterrence and provide access to MCISRE capabilities. As competition transitions to crisis, engineers provide mobility for augment forces and counter-mobility to deter or counter competitors. During transition from competition to conflict, material handling equipment at ports, as replacement for contracted companies, preparation work in anticipation for arrival of naval construction commands, force protection development for the JLEnt and other traditional base operating support-integrator responsibilities are critical for stand-in force consideration.

Missions that engineer units conduct to support the littoral force include but are not limited to: engineer reconnaissance; base camp or site preparation; detection and clearance of explosive hazards; airfield damage repair; horizontal and vertical construction; power generation; heavy equipment support; and bulk-fuel inventory management. Integration of naval engineering teams and capabilities in support of the littoral force can generate more robust and effective support, allowing for more rapid development of EABs. Under the engineering function, the following key tasks for the littoral force are particularly important in conducting EABO:

- Route and site reconnaissance to inform future dual-use and military construction projects
- Concealed and signature managed fuel distribution and storage
- Runway repair and preparation of potential FARPs
- Infrastructure improvement

6.3.4.1 Combat Engineering

Combat Engineering provides mobility, counter-mobility, survivability, and engineer reconnaissance to enable maneuver, protection, and sustainment of littoral forces. These tasks occur during all phases of operations and are integral to a littoral force's ability to maneuver. Combat engineers enhance the force's momentum by physically shaping the battlespace to support the survivability of friendly forces and make the most efficient use of the space and time necessary to generate speed while denying the enemy unencumbered maneuver. Engineer reconnaissance, including geospatial collections, enables greater awareness of micro terrain to inform operational planning in EAB site selection, mobility routes, LTPs, and preparing forward sites for aviation throughput. To support EABO, engineers will experiment with standard gap crossing capabilities and explosive hazard and mine detection and defeat capabilities to ensure mobility on land and at sea. Engineers will also conduct experimentation with land and sea mine capabilities to support sea control and sea denial tasks.

6.3.4.2 General Engineering

General engineering tasks include but are not limited to horizontal and vertical construction, bulk liquid operations, tactical water and/or hygiene services, bulk fuel operations, and tactical electric supply. Integration of naval engineering teams and capabilities in support of the littoral force can generate more robust and effective support, allowing for more rapid development of EABs.

Relying on organic capabilities of the littoral force to execute general engineering tasks typically requires significant transportation due to the size and weight of engineering and construction equipment and materiel. Transporting heavy equipment or material handling equipment creates a significant signature and is unavoidable. Local contracting may mitigate this concern, but OPSEC may be compromised in the process if additional efforts to hide intent are not included in planning and execution, especially in competition when multiple projects can occur to support future operations. The Theater Infrastructure Master Plan (TIMP) is where joint engineer projects are listed and prioritized. Including commander's priorities on the TIMP expands the engineer capabilities that may be applied to a project to include joint and contracted capabilities.

Current bulk fuel support methods mitigate risk by increasing days of supply levels to address resupply frequency challenges. However, while taking advantage of economy of scale, this creates valuable stationary targets. To sustain the future fight, a balance between days of supply levels stored and mobilized must be achieved. This can be done by focusing on caching, competition enabled fuel locations and signature-managed sites that limit targeting balancing the challenge of long-term caching with distribution and sustainment needs. Establishing and employing caches both afloat and ashore adds resilience to the distribution network, contributes to maximum availability, and improves flow of required resources. Ashore caching capability will require quick access to stocks at LTPs, taking into consideration quality assurance/surveillance and refueling mission (i.e., ground vs aviation). Caching and distributing fuel from commercial containers may permit the creation of ashore fuel farms without creating a significant signature. Similarly, the ability to procure locally available fuel or forage adversary or abandoned fuel sources and inject it with additives adds flexibility to the sustainment network, taking advantage of local resources.

6.3.4.3 Aviation Engineering

Engineers in the ACE provide limited combat and general engineering support to designated components of aviation operations, to include providing: engineer reconnaissance and survey; limited horizontal construction for the ACE (includes vertical or short takeoff and landing sites); constructing and maintaining mission-essential airfield requirements (temporary bunkers, aircraft revetments, and strongbacks); providing technical and equipment assistance for erection of pre-engineered buildings; providing utilities support (mobile electric power, water, potable water production, bath and laundry facilities, and refrigeration services); developing, improving, and maintaining drainage systems; providing technical assistance to support camouflage requirements; assessing bomb damage and providing airfield damage repair; providing material handling equipment services; and providing for EOD.

6.3.5 Explosive Ordnance Disposal (EOD)

Success in EABO will require integration of EOD personnel in direct and general support of maneuver, force protection, and intelligence. Considering the high threat and risk to force associated with mines, unexploded explosive ordnance, improvised explosive devices, enemy weapon systems, and weapons of mass destruction, EABO will rely upon EOD activities locate, engage, and defeat explosive hazards to personnel and operations through the very shallow water and surf zones, LTPs, and on land. EOD support requirements during EABO will be significantly higher than traditional combat operations due to the widely distributed nature of a contested littoral environment.

Due to the nature of EOD operations, EOD forces are frequently in the most advantageous position to collect actionable intelligence while performing their routine duties, leading to attribution, analysis of friendly and enemy capabilities, and ultimately, informing lethal effect decisions. Through reach back to various intelligence activities and the Joint EOD Technical Support Center, naval EOD forces can inject near-real-time information by providing immediate exploitation of collected exploitable materials.

6.3.6 Health Services

Health Services Support (HSS) minimizes the effects that wounds, injuries, and disease have on unit effectiveness and readiness. EABO requires an expeditionary HSS system that provides critical lifesaving care for longer periods of time, from the point of injury/illness to the next appropriate level of care by available transportation means, and an aggressive and proactive preventive medicine program that safeguards personnel. HSS to enable EABO against peer adversaries requires a complete paradigm shift from the Iraq/Afghanistan model (i.e., there likely will be no "golden hour" and minimal surgical capabilities). HNs will have medical systems that littoral forces could leverage; however, this capacity will likely be exceeded during armed conflict.

6.3.6.1 Principles of Health Service Support

The following six principles of HSS remain relevant in describing the overarching framework for EABO HSS planning, organizing, managing, and executing operations.
- Conformity: The medical plan must integrate and comply with the commander's plan.
- Proximity: HSS must provide the full spectrum of life saving care as close as the tactical situation permits to maximize the survival rate of wounded personnel.

- Flexibility: Expeditionary HSS must be prepared and empowered to shift medical resources to meet changing requirements, especially when casualty loads exceed treatment means and resources.
- Mobility: EABO will require medical assets to remain close to, or within supporting distance of, maneuvering forces, particularly when the OE is characterized by highly distributed operations and minimal, if any, pre-established health service HN infrastructure.
- Continuity: Achieving continuity of care is dependent on the ability to move casualties through progressive, phased roles of care, extending from the point of injury to definitive care. Each type of medical unit contributes a measured, logical increment of care appropriate to its location and capabilities. The EABO environment, with potential high casualty rates, extended distances, prolonged casualty holding, and high acute patient conditions, may necessitate that a patient receive treatment at each role of care or bypass roles as conditions allow.
- Coordination: Efficient employment of scarce medical resources is vital to the support of the tactical and strategic plan.

6.3.6.2 Health Service Support Planning

In EABO scenarios, HSS functions may need to provide casualty care and casualty holding for prolonged and uncertain durations. Considerations must be made when coordinating and employing HSS equipment sets and personnel to support prolonged casualty holding, casualty evacuation timelines, and medical logistics. Additionally, the distributed and highly mobile nature of EABO will require more consideration and evaluation of environmental threats such as water safety and infectious diseases. HSS units should be able to quickly determine environmental threats and risk to littoral forces. Coordinating HSS with HNs and capitalizing on HN capabilities may reduce the burden of providing this support organically and reduce the littoral force footprint ashore. Experimentation and exercising with both embarked and ashore HSS will help refine planning and TTPs for the provision of health services during distributed operations.

Several key tasks for the littoral force under the HSS function are particularly important in conducting EABO:
- Coordination of prolonged casualty holding periods and casualty holding and hospitalization sites
- Coordination across the naval force for evacuation of casualties and integration with theater patient movement systems
- Coordination and reconnaissance of HN, partner, and allied military and civilian health service facilities
- Management and distribution of shock-trauma and damage control resuscitation capability
- Management and distribution of surgical capability
- Coordination and forward positioning of Class VIII materiel

6.3.6.3 Medical Intelligence

Medical intelligence informs the expeditionary force commander and LFC of potential threats or hazards to military operations and is the basis of medical planning for EABO. It includes all-source intelligence on worldwide infectious diseases and environmental health risks, foreign military and civilian healthcare systems and infrastructure, and scientific and technical developments in biotechnology and biomedical subjects of military significance.

6.3.6.4 Medical Logistics Planning Factors

EABO medical logistics are tailorable to the mission, supported force, threat, and geography of the supported theater. It encompasses the functions of procurement, initial issue, materiel management, resupply, and disposition of material necessary to support naval expeditionary forces. Requisitions for Class VIII(A) materiel follow the same channels as other classes of supply. Under medical logistics, the following considerations are particularly important in support of EABO:

- Coordination across the naval force for evacuation of patients
- Liaison with the Theater Lead Agent for Medical Material (TLAMM)
- Items requiring special handling and storage
- Type and quantity of medical supplies needed
- Medical equipment maintenance and support requirements
- Blood product supply and distribution
- Specialized authorized medical allowance list (AMAL) configurations (e.g., Recon AMAL) for EABO forces

Health services support is reliant upon the specialized materiel and services provided by medical logistics; therefore, an integrated naval and joint theater informed medical logistics plan is an integral and critical component of supporting EABO. Medical materiel will typically flow through the same distribution channels and is subject to the same movement controls as all other classes of supply. To enhance EABO Class VIII(A) support, the Medical Logistics planner will:

- Identify the specified and implied time-phased materiel requirements necessary to support the operation plan (OPLAN)
- Identify the capabilities, limitations, and requirements of aerial and sea ports of debarkation
- Ensure coordination for the movement of supplies and equipment
- Identify forward positioned stocks in theater
- Identify HN support (HNS), if available
- Identify joint and multinational logistics support requirements to include the distribution plan

6.3.6.5 Medical Management Planning Factors

Conditions in EABO are seldom static and the flow of sick and wounded puts variable pressure on capabilities of medical sites. This environment requires a dynamic casualty management decision making process that must be applied at all levels within the medical management system. Clinical, logistical, and administrative services must maintain close coordination to achieve effective management of individual casualties. Above all, the basic objective is preserving life, limb, and eyesight.

Preventive Medicine: Preventive medicine seeks to avoid disease non-battle injuries from attriting the littoral force. This activity acts upon medical intelligence of the littoral operating area, assessing living conditions, ensuring adequate sanitation, and vector control among other tasks. Foraging operations will rely upon preventive medicine services to ensure the health and safety of the force.

Casualty Sorting (Triage): Littoral forces conducting EABO will sustain battle injuries and adversary long-range precision fires will likely create mass casualty events that will overwhelm locally available medical resources. In order to establish proper evacuation protocols, medical providers will conduct rapid evaluations to identify if patients are in need of immediate resuscitation, tolerant of delay in treatment, stable and capable of evacuation, or expectant.

Patient Holding: Disaggregated operations will require significantly more patient holding and movement capabilities due to time, distance, and medical evacuation availability due to capacity and/or adversary operations. Operationally dependent, the littoral force may employ HN facilities when casualties exceed organic medical resources.

Patient Evacuation: Casualties stable for movement should be evacuated as soon as practicable based on the availability of transportation (e.g., aeromedical, ground sources if HN capabilities exist, and medical ship) and tactical circumstances.

Patient Expiration: During armed conflict, forces executing EABO may not be able to evacuate casualties for life-saving care and patients will die of wounds during patient holding. HSS personnel will transfer expired patients to mortuary affairs personnel for dignified transfer. See MCRP 3-40G.3, *MTTP for Mortuary Affairs in Theaters of Operations*, for more information.

6.3.7 Services

Among CSS, civil affairs, mortuary affairs, law enforcement, real property management, and the CCF are particularly important in EABO. Civil affairs support underwrites the ability of the littoral force to gain and maintain HNS and to rely on HN services to optimize signature while maintaining requisite capabilities. Different than other deployments in the campaigning phase, Marines will live among the population, not removed from them. Sovereign countries take great risk and reap benefit from allowing Marine Corps commands in their borders. Striking the correct balance between "reach back" support to the continental United States (CONUS) and utilizing local services will entice the country to continue acceptance of U.S. troops. Services considerations are of the utmost importance for bridging this gap between our various countries.

The Naval Service must consider varied approaches to the provision of command services. For example, reach-back solutions to providing personnel administration and financial management services may be suitable, while religious ministries support likely requires a more direct approach in the context of EABO. Food service support may be best provided using an approach that combines OCS and naval logistics. Several key tasks for the littoral force under the services function are particularly important in conducting EABO:
- Leveraging commercial and HN capabilities
- Contracting services and materiel
- Disbursing and postal services statutory requirements will require unique solutions given challenges associated with distributed operations
- Develop a base camp plan which incorporates real property management considerations for work and living spaces
- A robust administrative and legal apparatus is required to ensure naval personnel's legal rights are maintained
- Increased local law enforcement education is required for EABs. Units rotating into an unfamiliar LOA will experience unfamiliar local ordinances and laws which increases the probability of violations

6.3.7.1 Mortuary Affairs

Facing a peer adversary, mortuary affairs support faces significant challenges during the conduct of EABO with forces operating in a highly mobile and dispersed manner throughout the littorals. Lethal attacks may result in fatality management that outstrip the joint force's capacity to handle remains. The LFC must coordinate with the joint mortuary affairs command to ensure this service is provided to fallen Marines and Sailors.

The Marine Littoral Regiment (MLR) must consider having fatality management personnel on their staff, a collateral billet filled by a 0407 MOS[43]. Additionally, there should be a search and recovery (S&R) team assigned to each element of the MLR or stand-in force to meet the needs of mass casualties until the Personnel Retrieval and Processing (PRP) Company is activated and in theater or a joint mortuary affairs collection point is established. Specific training is required for the S&R team to ensure dignified handling of all remains, personal property retrieval, and appropriate processing to the collection point. Temporary interment and disinterment require CCDR approval before executing.[44]

Mortuary affairs activities include:
- Search and recovery missions
- Operate mortuary affairs collection points
- Conduct temporary internment and disinterment
- Operate theater mortuary evacuation points

6.3.8 Aviation Logistics

Marine aviation logistics (AvLog) is a critical element to successful ACE operations, especially in EABO. AvLog is a discrete set of sustainment activities in direct support of the ACE, centered on authorized repair of aircraft executed by Marine maintainers, and further enabled by an extended Naval Aviation Enterprise (NAE) - inclusive of Navy commands within the Navy's Supporting Establishment. As designed, AvLog is responsive to consumption and replenishment of various classes of supply and commodities necessary to maintain and operate ACE platforms. While integrated into the theater logistics posture, aviation sustainment professionals do not retain a core requirement to employ conventional ground logistics support. The ACE is traditionally viewed as self-sustaining, limiting the understanding and ability to influence sustainment outcomes by conventional logistics managers. AvLog remains the exclusive sustainment arm of the ACE in all operations.

The activities associated with AvLog are multi-echeloned and scalable to meet sustainment needs across a designated area of operations. Encompassing the functional areas of maintenance, supply, ordnance, avionics, information systems, and data administration, AvLog requires varying degrees of infrastructure to perform associated tasks. These infrastructure requirements can be tailored to support remote or expeditionary settings within acceptable tolerances and safety-based, prescriptive parameters. Pre-positioning and redistribution occur based on both scheduled and unscheduled sustainment need, but do not fundamentally alter the regimented approach to sustainment execution. A broader MAGTF logistics concept of support continues to be defined with specific attributes that enable survivability and resilience in what is assumed to be a less permissive and contested environment.

As a universal logistics consideration, more extreme geography, distance, and the enemy's denial of concentration to include larger scale logistical hubs must be addressed. In certain functions, the future of scalable aviation sustainment may look as distributed as the aviation forces supported. While remaining in compliance with the Naval Aviation Maintenance Program, an episodic framework consisting of existing and future enabling platforms could be leveraged to return aircraft to mission capable status. Further enabled by more modern digital tools, the AvLog ecosystem may become even more integrated – across the Service, Naval, and Joint force – with the implementation of artificial intelligence/machine learning as a true force multiplier that facilitates a more dispersed, prognostic system of aviation support. Continued analysis and experimentation is required to better identify the associated capabilities and attributes necessary to support DAO.

---

[43] 0407 MOS is obtained through the Army's two-week Mortuary Affairs Officer Course, course code 8B-SI4V.
[44] Headquarters, US Marine Corps, Multi-Service TTPs for Mortuary Affairs in Theaters of Operations, MCRP 3-40G.3 (Washington, DC: US Marine Corps, 2022)

The following areas must be considered when planning for EABO:
- Alternative methods of obtaining fuel will be necessary. ACE planners should expect to use fuel from many sources, including host nations, portable fuel bladders, and pre-positioned stocks. There must be less emphasis on using motor transport assets to move fuel.
- Marine Corps aviation maintenance and supply functions must seek even greater integration with the naval aviation logistics enterprise to fully leverage the complementary aspects of ashore and afloat aviation maintenance and supply capabilities. New methods of delivering maintenance equipment, spare parts, and technicians must be explored so that aviation maintenance capabilities can be distributed across the WEZ to complicate adversary targeting. Temporary aviation maintenance locations will need to be established for short time periods to conduct specific maintenance functions. Other maintenance functions should be located outside the adversary WEZ.

6.3.9 Safety

Planning for safety during EABO encompasses a number of functions that occur at all levels of command, and the integration of risk management controls is critical. Coordinating safety support and capitalizing on collateral duty capabilities may reduce the burden of providing this support organically. Experimentation and exercising with both embarked and ashore safety support will help refine planning and TTPs for the provision of risk management and safety during distributed operations. Safety considerations should include geospatial and weather data, sea states, beach and surf zone hazards, and special equipment hazards. Planners must utilize mishap and safety recommendations and lessons learned.

6.4 OPERATIONAL-LEVEL LOGISTICS

In the Marine Corps, operational-level logistics orients on force closure, arrival and assembly, intra-theater lift, theater distribution, sustainment, and reconstitution and redeployment. These core elements require the art and science of utilizing joint, Service, and strategic-, operational-, and theater-level enablers to support and sustain a tactical FMF. To be successful at this level, the Marine Corps component command operational-level planners must understand Service, HN, JLEnt, coalition, NGO, private volunteer organizations, Department of State (DOS), and theater logistical capabilities and know when and how to request their support. Each Marine Corps component command is responsible for design, planning, coordination, and execution of operational-level logistics for its geographic area. The Marine Corps component commander is therefore required to devise appropriate operational-level logistical support and coordinate with those Marine Corps organizations that support applicable joint and theater enablers to execute the CCDR's OPLANs and contingency plans.

6.4.1 Force Closure

Force closure as a logistics function is the calculated accumulation of a unit or capability to accomplish a designated mission. This can be considered the "integration" capability tied to Reception, Staging, Onward Movement, and Integration (RSO&I), particularly when the integration occurs away from a port of debarkation. Force closure is specific to a unit or capability, although a mature operating theater may have elements assigned specifically to focus on rapid deployment and employment of disparate force elements to ensure the right capability is ready at the right time and in the right place. Of note, equipment capabilities globally positioned forward in a theater of operations enable more rapid force closure by reducing the requirement for global, and possibly intra-theater lift.

6.4.2 Arrival & Assembly

Forces conducting EABO must be capable of deploying with no or minimal arrival and assembly support to enable the naval SOM. During operations, this arrival and assembly activity supports both tactical maneuver and sustainment activities and is supported by organic elements and the Pioneer Battalion (described in sections 6.4.3.4 and 6.4.3.5).

The LFC must consider many factors, such as:
- Organic and/or attached intra-theater movement platforms available to close the littoral operating area
- Distance between present location, home station, and the assigned operating area
- Port of debarkation facilities, to include previously tactical locations such as landing beaches and landing zones, within the assigned operating area capable of receiving intra-theater movement platforms
- Requirements for arrival and assembly support
- Adversary interdiction capabilities during force closure
- Command relationships of all forces involved

6.4.3 Intra-theater Lift

Intra-theater lift is the broad array of conveyance assets in a theater of operations that are accessible for forces in a particular theater for distribution, sustainment, and potentially operational maneuver. This often includes airlift and sealift capability assigned to or apportioned to a particular theater of operation or campaign. In large land operations, this may also include rail capability for continental transportation.

A range of surface connectors and multi-capable distribution platforms are needed to facilitate this requirement. Intra-theater lift typically includes military assets, as well as commercial lift capability and capacity that may be part of standing or emergent contracts or HN support. EABO planners should analyze transportation methods utilized by local national civilians and businesses to identify what forms of conveyance work best in a given LOA. Considerations of other sources for this capability needs to be further researched and experimented with to meet the needs of an EABO mission.

6.4.3.1 Littoral Movement

Force closure, maneuver, and sustainment of naval forces within contested areas are essential to the success of the littoral force. Employment of numerous, small, versatile transportation assets permits naval logistics to disperse, enables maneuver and mobility, and provides resilience across the force. Persisting inside the WEZ requires frequent maneuver of forces along the littorals to achieve positions of advantage relative to adversary capabilities and enable survivability and sustainment. Naval forces executing EABO must be able to quickly maneuver over operational and strategic distances directly to tactical assembly areas/operating areas with minimal RSO&I. This requires multimodal transportation solutions and leveraging prepositioning when possible.

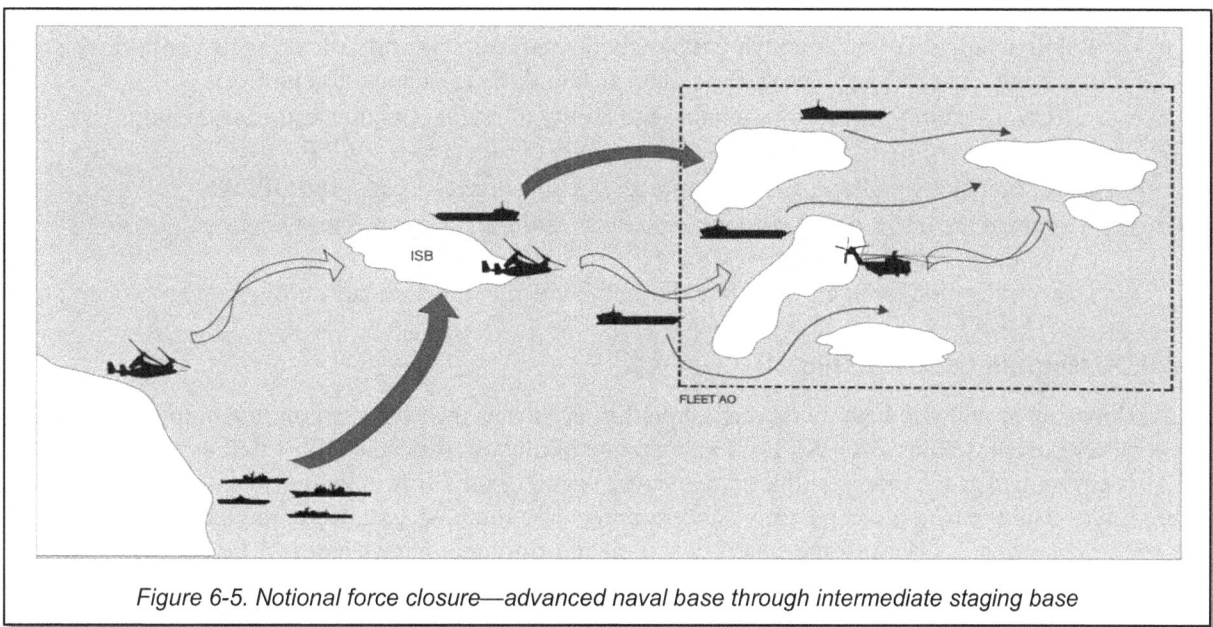

Figure 6-5. Notional force closure—advanced naval base through intermediate staging base

Once established in the LOA, naval forces must maneuver tactically within the littoral, including (1) inter-island movement within an archipelago, (2) intra-island movement via ground, surface, or air modes, and (3) even displacement to another LOA. This maneuver allows the naval force to gain advantage by occupying or controlling key maritime terrain, remain survivable despite adversary targeting attempts, and execute or contribute to deception operations.

Lastly, littoral maneuver assets must contribute to the sustainment of naval forces operating within the LOA. These assets must be able to provide area and delivery support across the beach or through the air to augment landward logistics networks. When performing these functions, these afloat maneuver assets serve as seaward nodes for the logistics network supporting an EABO.

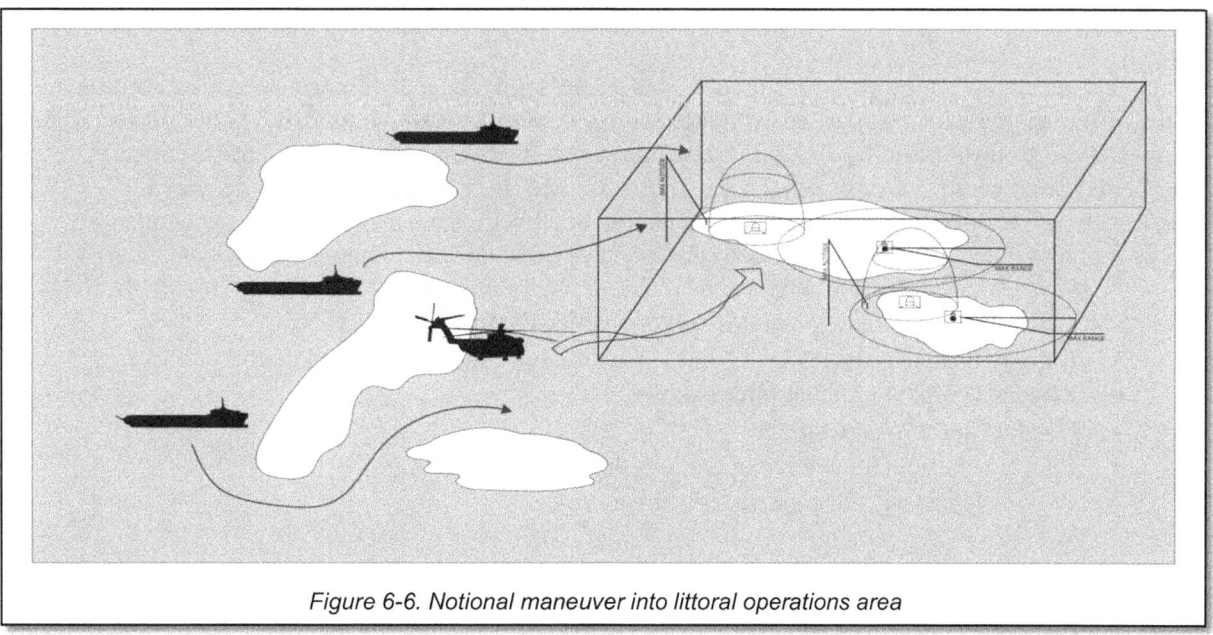

Figure 6-6. Notional maneuver into littoral operations area

The littoral force's ability to maneuver and operate within the littorals generates the following benefits:
- Rapid deployment and employment based on a reduced footprint ashore while configured to consume less and be able to operate in areas with little to no infrastructure
- Agility, capability to rapidly displace, and ability to retain effectiveness while mobile
- Persistence, ability to survive and endure within the adversary's WEZ
- Deception, not simply providing a service, but an active element of an OPLAN
- Regeneration, derived from no single points of failure, no linear lines of supply, and a "honeycomb" distribution system
- Decentralization, with logistics functions embedded in all elements during maneuver operations

6.4.3.2 Medium Landing Ship

Littoral maneuver will rely heavily on surface platforms such as the Medium Landing Ship (LSM), Next-Generation Logistics Ships (NGLS), LSM and a range of other multi-capable distribution platforms. The LSM is envisioned as the principal littoral maneuver vessel of the MLR. Command and organization of these forces could remain under a purely Navy command or could be assigned to the FMF as previously described in section 2.5. Live-force experimentation and wargaming must assess different organization and command arrangements.

6.4.3.3 Medium Landing Ship Employment

The LSM and NGLS support the day-to-day maneuver of stand-in forces operating in the LOA. It complements larger amphibious warfare ships and other surface connectors. Utilizing the LSM to transport forces reduces the impacts of tactical vehicles on the road network, increases deception, and allows for the sustainment of forces during embarkation. The range, endurance, and austere access of LSMs enable the littoral force to deliver personnel, equipment, and sustainment across a widely distributed area.

Shallow draft and beaching capability are keys to providing the volume and agility to maneuver the required capabilities to key maritime terrain. LSM employment requires reconnaissance and prior planning relating to the bathymetry of the littoral environment. Effective LSM employment relies on knowledge of the beach makeup, slope, currents, tidal effects, and other environment factors.

As envisioned and when properly postured, LSMs possess the range, endurance, speed, seakeeping, and C4ISR capabilities to support and conduct complementary operations with, but not as part of, U.S. Navy tactical groups, including an ESG or amphibious ready group (ARG). Forward-positioned LSMs may augment the capabilities of deploying ARG/MEUs during regional engagement and response to crises or contingencies. Experimentation must help planners understand the competing needs for rapid maneuver and the ability to provide tactical level sustainment from the sea.

The LSM with embarked forces, generates and/or enables the following effects:
- Rapidly maneuver forces from shore-to-shore in a contested environment
- Sustain a combat-credible force ashore
- Conduct enduring operations
- Enable persistent joint-force operations and power projection
- Provide increased and capable forward presence

6.4.3.4 Pioneer Battalion

The Pioneer Battalion operates across the competition continuum. During competition, Pioneer Battalion enables reconnaissance – counter reconnaissance and posture efforts. In crisis and armed conflict,

Pioneer Battalion conducts assured mobility across seaward and landward portions of the littoral operating area in order to enable littoral force maneuver and persistence in a contested environment while supporting sea denial operations as part of the Naval Expeditionary Force. The Pioneer Battalion employs a hybrid team of naval capabilities unconstrained by the water depth, reducing obstacles and improving routes from the very shallow water limit, inland to interior lines of communication. Pioneer Battalion's engineers and littoral explosive ordnance neutralization capability complements the Navy expeditionary mine countermeasures capability. Capacity is sufficient to support concurrent employment of the surface connectors and vessels of a Littoral Maneuver Squadron (LMS) sustaining three MLRs deployed in the littoral operating area.

The Pioneer Battalion's mission-essential task list includes:
- Conduct mobility operations
- Conduct countermobility operations
- Conduct general engineering
- Conduct forward aviation combat engineering
- Conduct engineer reconnaissance
- Conduct landing support operations
- Conduct EOD operations
- Conduct bulk fuel operations
- Construct and improve LTPs
- Conduct survivability operations

6.4.3.5 Pioneer Battalion Employment

The Pioneer Battalion is best employed as a general support unit during campaigning. Transitioning to conflict, Pioneer Battalion is best employed as a direct support asset to the Division and employs three purpose-built Pioneer Companies, one company in direct support to each MLR deployed as part of the stand-in force. Each company has sufficient capability/capacity to plan, prepare, operate, and provide assured mobility from LTPs landward. Limited Marine general engineering capabilities are enhanced by the multi-functional Naval Mobile Construction Company through vertical and horizontal construction and underwater construction teams. The Pioneer Battalion capitalizes on presence in the littoral operating area during the campaigning phase through subject matter expert exchanges, TSC exercises, and humanitarian assistance activities.

6.4.4 Theater Distribution

A supply chain network, or supply web, is an engineered flow of information, funding, or materiel from suppliers to tactical customers. Theater distribution is the process that coordinates, synchronizes, and prioritizes fulfillment of requirements from point-of-debarkation in each theater to the point-of-need. The intent of this process is to synchronize all elements of the logistics system with the distribution network to deliver the right things to the right place at the right time. Organizations provide data on the real-time status of supplies and suppliers so logisticians can leverage the JLEnt and adjust as necessary to a dynamic and changing operating environment.

6.4.5 Sustainment

Sustainment for EABO requires the arrangement of key logistics personnel and logistical resources at the right place within a theater to extend the reach of the force, particularly in time. This differs from tactical logistics, in that the essence of sustainment as a function is to enable the functioning of military capability across an entire theater of operations. Like force closure, sustainment was previously a function of logistics conducted at the component HQ-level and coordinated at the task force-level. The same intra-theater movement platforms and arrival & assembly capabilities that enable force closure may also sustain littoral forces concurrent with tactical/operational maneuver. EABO forces will draw sustainment via organic and assigned aircraft and ships from afloat forward logistics bases and shore-based cache sites outside of their EAB and the overall LOA. Moreover, theater distribution forces will flatten the supply web and deliver sustainment directly to tactical forces conducting EABO, bypassing intermediary logistical commands and establishing a "honeycomb" of logistical support sources.

The LFC must consider many factors, such as:
- Distance between EABs, afloat logistics support bases, and advance naval bases
- Location of and access to afloat and ashore globally positioned materiel
- Locations and disposition of ashore, afloat, subterranean, and underwater cache sites and local sources of supply
- HN contractor vetting processes and approved sources of supply
- Organic and/or attached intra-theater movement platforms available to conduct sustainment and distribution
- Throughput capabilities and capacities within the LOA
- Enemy interdiction capabilities
- Command relationships of the sustainment network

6.4.6 Reconstitution & Redeployment

Because expeditions are, by definition, temporary, all expeditionary operations involve a redeployment of the expeditionary force or a transition to a permanent presence of some sort. Reconstitution and redeployment are combined to account for the activities conducted in theater to consolidate capability and reorient for follow-on tasks. Reconstitution includes activities to recondition forces to a necessary level of operational value appropriate with mission requirements and available resources. Effective reconstitution and redeployment is not as simple as a tactical withdrawal of the expeditionary forces from the scene and reemploying that force on new terrain. It requires withdrawing the force in a way that maintains the desired situation while preserving combat capabilities of the force. For example, care must be taken to reload ships of an MPF or MEU to restore their sustainment capabilities because either force may be instantly ordered to undertake another expeditionary operation. US Transportation Command (USTRANSCOM) and JLEnt agencies participate heavily in the redeployment of forces.

6.4.7 Component and Operational-Level Logistics

The naval force must operate under a single, coordinated, and integrated naval logistics architecture capable of addressing supply and maintenance issues for all systems from warships to rifles within the LOA. Ideally, this takes place through a single automated information system that is ubiquitous, accessible, and secure. At the operational level of warfare, the fleet will sustain the littoral force from distributed bases across a theater directly alongside Naval Expeditionary Combat Force elements, US Coast Guard forces, and the joint force to ensure flexible and resilient capability and capacity.

The Marine Forces component is the responsible headquarters, under Title 10, for the sustainment of Marine units operating within the AOR. It coordinates the fulfillment of Marine Corps logistical requirements on behalf of the Commandant of the Marine Corps to the CCDR, JFC, and the Naval Force component / major / numbered fleet commander under whom Marine forces are operating. Specific to Marine Forces Pacific (MARFORPAC), the Fleet Marine Force Logistics Command (FMFLC) is envisioned as a standing operational logistics headquarters to expand the capability and capacity of the MARFORPAC G-4 staff, fulfill JFMCC N4 staff requirements, and provide a deployment distribution operations center capability to U.S. Pacific Fleet. FMFLC can provide command and control over

Marine forces operating advanced naval bases across the naval sustainment network in both competition and conflict and serves as a conduit to the Marine Corps Logistics Command.

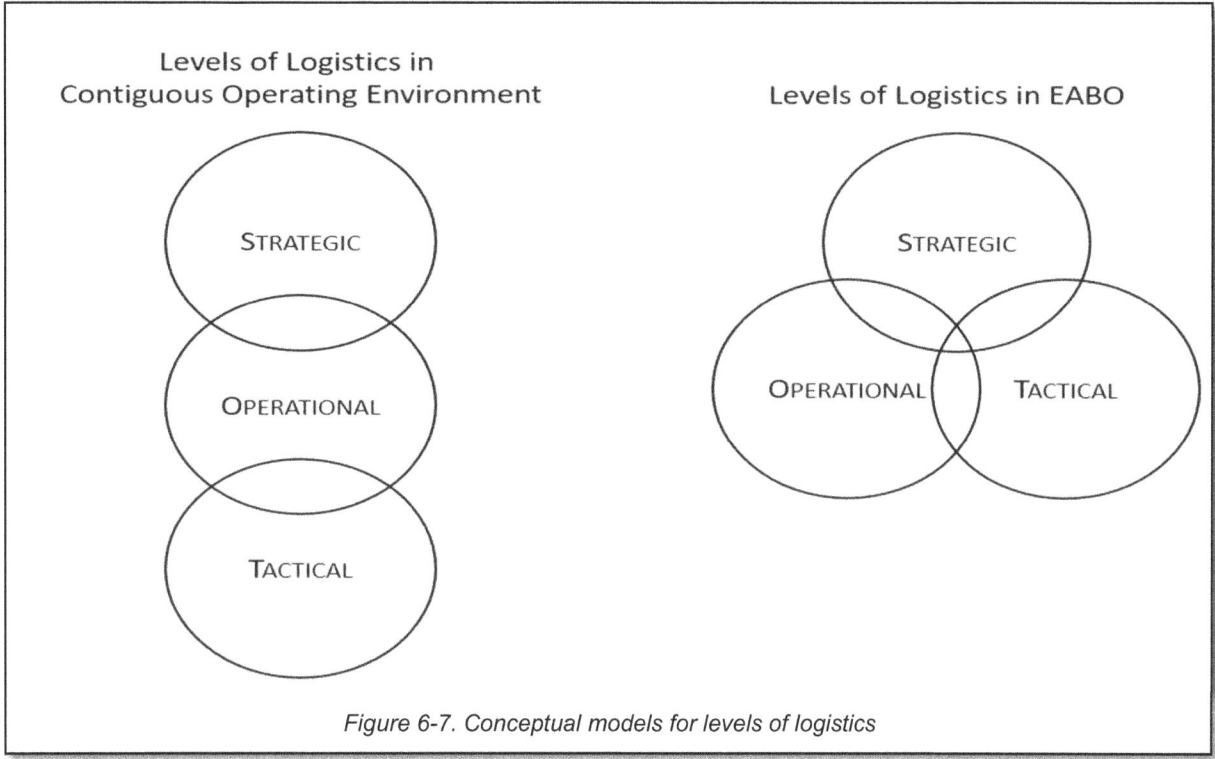

Figure 6-7. Conceptual models for levels of logistics

6.5 STRATEGIC-LEVEL LOGISTICS

Strategic-level logistics encompasses the nation's ability to raise, deploy, and sustain operating forces in the execution of the national military strategy. It supports organizing, training, and equipping the forces that further national interests. To conduct successful sustainment of littoral forces requires the compression of the levels of logistics: strategic, operational, and tactical. Linear systems do not provide the littoral force with the resiliency required to persist while conducting EABO. New technology and approaches (e.g., additive manufacturing and operational contracting support) postures the strategic defense industrial base forward at the tactical level in the hands of the warfighter.[45] Due to the complexity of this integrated construct as depicted in figure 6-7, sustainment normally occurs at the task-force level and higher.

While the task-organized littoral force, at the task-group level and below, places primary emphasis on the six functional areas of tactical-level logistics, littoral force logisticians must also consider strategic- and operational-level functions of logistics. Logistical planners must know and understand all levels of logistics (tactical, operational, strategic), including those capabilities resident in the JLEnt and HNs. They must be skilled with integrating these capabilities and functions with the broader operational plan. These would include the U.S. Embassy, United Nations, coalition forces, ongoing DOS missions, and possibly programs being executed by the U.S. Agency for International Development to name a few. Of particular importance is the understanding of the capabilities tied to littoral distribution to ensure logistics activities do not compromise operational maneuver.

---

[45] Headquarters, US Marine Corps, Contingency Contracting, MCRP 3-40B.3 (Washington, DC: US Marine Corps, 2018)

The eight functions of logistics at the strategic-level are mobilization, procurement, war reserves, facilities, material readiness, strategic airlift and sealift, deployment and support, and force regeneration. All these functions will support the LOA and deployed force directly or indirectly during the competition continuum.

6.5.1 Mobilization

As the competition transitions to crisis or contingency, mobilization of additional forces may be appropriate to augment the MLR or stand-in forces. An EABO environment may require mobilizing and activating of specific capabilities only found in the Reserve Component (i.e., law enforcement, PRP and civil affairs personnel.) This may include units not yet ready to deploy on short notice. Mobilization requires specific authorities for activation, although this may be appropriate for specific crisis or contingency response. The commitment of designated forces and capabilities is a Service decision, and therefore the management of those resource areas is typically conducted at the Service level.

6.5.2 Procurement

Typically, procurement is a long-term acquisition process for weapon systems to meet operational missions. Due to the lengthy process for procurement of equipment with military specifications, forces should seek options that are commercial off-the-shelf to expedite acquisition and to leverage existing commercial supply chains for system sustainment. Given the unique OE of LOA, forces may create and submit Urgent Universal Need Statement requirements for unforeseen equipment and capabilities. Traditional procurement will not be responsive enough to respond to a force conducting EABO, although through informed procurement, the force conducting EABO will eventually be sourced with the right equipment.

6.5.3 War Reserves

War reserve materiel (WRM) supplements peacetime operating stocks to meet the total USMC requirement of equipment and supplies to train, equip, field, and sustain forces during wartime. The Marine Corps invests in WRM to ensure the Service has on hand or has the ability to rapidly acquire the materiel needed to resupply deployed forces before theater logistics begins delivering common item support. Procured WRM will be prepositioned as far forward as possible to improve response time as part of the materiel risk management plan. WRM includes afloat and ashore globally prepositioned materiel and stocks held in stores to supplement Stand-in Force organic capabilities.

6.5.4 Facilities

In leveraging Service-level logistics systems, Navy and Marine Corps striking power is sustainable through an established network of bases and installations. To enable EABO, forces will tie back through operational-level logistics to bases and stations that will support force closure, force projection, and force sustainment. Each base must be considered a force projection platform as part of a resilient network that is part of theater and global posture. Investment in facilities during campaigning, including protection and resiliency, assures partners and allies, and has the potential to deter potential adversaries. These CONUS and advanced based facilities are an integral part of the sustainment effort, and the littoral force must understand how to leverage their capabilities within the JLEnt to provide sustainment to an EABO.

6.5.5 Material Readiness

Materiel readiness as a strategic-level logistics function relates to the enterprise lifecycle of equipment across Service. Related to EABO, the Service will maintain awareness of maintenance and readiness of equipment used to conduct EABO operations. Through coordination across Service equities, equipment will be managed to assure high states of readiness, as well as the most updated and appropriately modified equipment be ready where and when the equipment is required to meet a mission.

6.5.6 Strategic Sealift & Airlift

Strategic lift is the bridge for forces to flow from outside of a theater into the theater of operations. While responsive in crisis and contingency, coordination during campaigning requires prioritization across concurrent global demands. Assumptions regarding the operating environment, will limit the use of strategic lift to sustain EABO during crisis and contingency response. Planners will need to plan how strategic lift will be utilized in order to plan the deployment and sustainment of forces into the EABO AO. Serving as global distribution coordinator, USTRANSCOM is responsible for establishing processes to plan, apportion, allocate, route, schedule, validate priorities, track movements, and redirect forces and supplies per supported commander's intent. These responsibilities are outlined in the unified command plan, on behalf of and in coordination with the joint deployment and distribution enterprise community of interest.

6.5.7 Deployment & Support

Deployment of forces for EABO in competition is likely to be routine. During crisis and contingency, forces that are generated and deployed from home station will compete with the force flow of many elements from across the joint force and from many different locations. For this reason, forces deploying in support of EABO should be prioritized as they are likely to arrive later than the situation warrants. Commanders can condense deployment response by pre-staging tailored flexible response option force elements.

6.5.8 Force Regeneration

Force regeneration relates to the restoration of units to a desired level of combat effectiveness commensurate with mission requirements and available resources (JP 3-02). As forces in EABO may be attrited, it will be necessary to replace capabilities to regenerate combat effectiveness. Combat replacements (material, equipment and personnel) in theater will be limited, although prioritization, allocation, and apportionment of critical capabilities may result in relatively rapid regeneration to support critical capabilities. In other cases, force regeneration will be delayed with reliance on capabilities that are outside the theater. Force regeneration in theater will be a focused effort by GCC and Service component commands. This can be mitigated with some theater reserve capabilities or specific replacement elements at a high state of readiness outside of the theater.

6.6 CONCLUSION

Sustaining the force through all its missions in the competition continuum, around the world, and in crisis or conflict is a complex endeavor. Given the idiosyncrasies of distributed maritime and expeditionary advanced base operations they are compounded even more. The logistics community must understand and capitalize on the JLEnt, other sources of sustainment, and be subject matter experts on the 20 functional areas of Marine Corps logistics at the strategic-, operational-, and tactical-level to mitigate future requirements and fulfill this essential warfighting function. Education, experimentation, acquisition, and training is the recipe to advance the logistics community to ensure our Service meets the challenge of the next conflict.

INTENTIONALLY BLANK

CHAPTER 7

LITTORAL OPERATIONS

7.1 GENERAL

This chapter serves as the tactical and operational construct for planning EABO in support of littoral operations and conducting live-force experimentation and wargaming. It outlines considerations for preparing the OPLAN and addresses unique factors specific to mission sets and operations conducted by the littoral force. Essential to preparation of the OPLAN is a concept of operations that views activities as an integrated whole within the maritime domain. Critical planning considerations include application of warfighting functions to task organize elements capable of supporting warfare commanders with critical capabilities. Littoral operations are not merely reactive to crisis; rather, they validate the baseline of joint-force activities in day-to-day campaigning and provide a deterrent capability to contain adversary aggression in escalation.

7.2 CONCEPT OF OPERATIONS

The development of the concept of operations is an iterative process. Detailed planning refines the initial concept and is an extension of day-to-day activities in the operating area. Critical to the success of units conducting EABO is a concept of operations that deliberately manages the OE by accounting for all activities across the competition continuum. The concept of operations will include deliberate condition setting through coordination with the DOS, HN agencies, and military partners as discussed in chapter 2. Several factors may necessitate modifying the concept. They include operational requirements of littoral force elements, shifts in the OE, changes to access/basing/overflight permissions, and changes in the adversary posture. Throughout its formulation, the concept provides the basis for detailed and concurrent planning and is included in the OPLAN to clarify the commander's purpose.

The detailed concept of operations outlines the commander's COA decisions and visualizes the operation. It depicts a broad outline of the plan of execution. At a minimum, it should include the purpose and scope of the operation, major or essential tasks, and phasing or sequencing of actions to shape and assess events across the competition continuum.

7.3 PLAN OF EXECUTION

The plan of execution provides for the employment of the various elements of the littoral force. It consists of three parts:

<u>Scheme of Maneuver</u>. Description of how arrayed forces will accomplish the commander's intent. It is the central expression of the concept for operations and governs the design of supporting plans or annexes. Subordinate documents to the SOM include the littoral maneuver plan and the plan for supporting operations.

<u>Littoral Maneuver Plan</u>. This plan covers the seaward and landward maneuver of forces to and within the LOA. Methods of maneuver may include any combination of amphibious warfare ships, NGLS, LSMs, and surface connectors and craft (both manned and unmanned), as well as ground and aviation assets. The littoral maneuver plan must thoroughly address transitions between domains and multimodal transportation methods. The organization of a LMS comprising various shipping options assigned to task-organized littoral forces will be developed during initial planning as discussed in chapter 2.

Plan of Supporting Operations. The elements of the plan of supporting operations shape and establish conditions for executing the SOM and accomplishing the mission. Elements of supporting operations may be delineated according to warfighting function, domain, and/or civil-military considerations. The following paragraphs, while not all encompassing, serve as a baseline for planning supporting operations.
- Plan for Information. Actions in the OE, such as the physical movement and activities of combat systems and personnel, generate effects in both the OE and the IE. Consequently, information activities should be planned and executed to both enable the littoral force's scheme of maneuver and leverage maneuver activities that impact the IE. Actions in the OE impact the IE and must align to and support the JFC's efforts, especially those meant to inform, influence, and deceive target audiences. Failure to plan for and align these efforts may produce a "say-do" gap that creates a potential vulnerability for the adversary to exploit or results in a loss of trust with partners and allies.
- Host Nation Coordination. Mission requirements influence the spectrum of HN coordination options available to planners. Options include contingency contracting and HNS. HNS may include preplanned contracting effective in time of conflict, which is known as wartime HNS. Plans must identify the authorities necessary to execute pre-arranged agreements and contracts in support of the littoral force. Contingency contracting officer placement within appropriate forward elements of the littoral force is an essential element of contracting support.
- Reconnaissance. Reconnaissance is a mission undertaken to gain information about the enemy and the meteorological, hydrographic, or geographic characteristics of a particular area. Further, obtaining information about the activities and resources of potential adversaries, local populations, and other related groups provides the littoral force with baseline situational awareness prior to mission execution. Reconnaissance is a continuing multidomain action in EABO. Because dispersed and highly mobile littoral forces will frequently displace, continuous assessment of the OE is essential to the commander's planning to maintain a current and accurate picture of the displacement routes and destinations.

7.3.1 Plan for Sustainment

The concept of support establishes the plan for logistics and sustainment of the littoral force from embarkation through termination of littoral operations. This plan accounts for the employment, synchronization, and coordination of the littoral force's logistic elements, afloat fleet logistic task forces, and component and theater-level logistic commands.

7.3.2 Plan for Aviation

The plan for aviation support to the littoral force is guided by command relationships and the joint or combined force commander's aviation apportionment and allocation decisions. This plan coordinates the activities of organic littoral force aviation with combined and joint assets. Air operations executed by naval expeditionary force air elements and other joint air assets complement one another and constitute a collective capability supporting the concept of littoral operations. Littoral force aviation must be able to configure their combat systems from relatively secure areas, move into operational positions, set up and operate for limited duration missions, relocate within the enemy's targeting cycle, and then repeat as necessary. Sequential operations from split sites may be necessary to accomplish longer duration missions.

The following are considerations for the employment of units of the littoral force ACE:
- Air-control agencies, call signs, and frequencies
- Air missions in support of task-organized littoral forces (e.g., OAS, assault support, aerial reconnaissance, and EW)

- Employment of organic antiair and counter-UAS capabilities
- Joint targeting processes and making use of the joint ATO
- Air control measures and fire support coordination measures
- Aviation logistics support
- Aviation ground support
- Air support, assault support, and medical evacuation request processes

7.4 COMMON PHASING CONSIDERATIONS

Littoral operations may take many forms and require various task organizations of the littoral force. Activities in assigned operating areas and within LOAs may vary widely depending on assigned missions and tasks. However, movement to and occupation of designated localities for EABs have common phasing considerations across the range of available mission sets. The following paragraphs discuss common considerations that address the mission planning necessary to conduct littoral operations.

7.4.1 Shaping and Reconnaissance

As discussed in chapter 2, the littoral force conducts continuous shaping activities at echelon to set the conditions for littoral operations. These actions set conditions for the various task elements throughout the developed EAB(s) to gain access, occupy, and employ capabilities within assigned positions. Prior to final force closure, reconnaissance activities validate existing gaps in information and inform planning. The LFC may consider the following prior to and during EAB occupation:

- Conducting detailed intelligence preparation of the battlespace (IPB) of assigned battlespace
- Adversary posture and levels of activity
- Surveys of LTPs including: airfields, runways, landing sites, and landing points to determine suitability
- Establishing baseline of EMS
- Conduct counter-C5ISRT
- Proximity to key maritime terrain where forces will be delivering effects
- Mobility in and out of LOA(s), positions, and hide sites
- Natural concealment available in the terrain
- Establishing pattern of life/local activities
- Verifying partnered force/HN training goals

The LFC must leverage support from MARFORSOC and other SOF elements and their relationships, capabilities, and authorities in the operating area. Making liaison and establishing linkages with SOF elements during detailed planning facilitates expanded access to pre-established and mature relationships with agency partners, coalition partners, and local state and non-state partners.

7.4.2 Position Selection and Improvement

LFCs will need to select several primary locations for employing their capabilities along with multiple alternate and supplemental positions in any operation. They must balance competing considerations of system ranges, proximity to vital areas and key maritime terrain, and threat levels when selecting the best positions for mission accomplishment. Activities to prepare and improve operational positions and locations by TSC initiatives, an advance party, or MARFORSOC elements prior to littoral force arrival are key. Preparation of physical locations helps generate relative tempo as the littoral force is in transition during the conduct of littoral maneuver. The littoral force must continually reassess and improve positions relative to the changing OE and adversary threat posture. Key considerations for position selection include:

- Validating infrastructure surveys

- Liaison with appropriate level HN governmental authorities
- Liaison with local contracting officials
- Assessing local population atmospherics
- Proximity to HN forces and civilian population
- Proximity to key infrastructure—for example, runways and port facilities
- Establishing decoy units
- Prestaging of caches
- Availability of supplementary and alternate positions

7.4.3 Occupying the Expeditionary Advanced Base

This section provides information to littoral force commanders on the organization, control, and execution of movement to an assigned LOA and occupation of an EAB as outlined in the SOM and littoral maneuver plan.

Planning Considerations. Sequencing the establishment of capabilities and assessing levels of security are key considerations in arraying the desired in individual sites to be occupied. Experimentation may reveal that EABO, akin to forcible entry, may consist of a rapid buildup of combat power (including sensing, queuing, force protection, network integration), whether relying on swift occupation of a defensive position or resorting to techniques for infiltration. Planning aims to ensure forces, equipment, and supplies land via sea and/or air at the prescribed times and locations and in the operational posture required by the SOM.

Basic Considerations. The requirements for support to initial EAB loading are (1) preserving tactical integrity of the littoral force and (2) achieving optimal dispersion of forces and assets used in the littoral maneuver plan.
- Tactical Integrity. The organization for loading an EAB must assure adequate control of dispersed littoral forces while retaining overall tactical control by subordinate commanders within an EAB. The force preserves tactical integrity by combat spread loading of LMS shipping and littoral force aviation assets. The tactical integrity of littoral force elements does not require an entire element to embark on a single mode of transportation. For example, a unit of action assigned to conduct SUW may embark in several LSMs while littoral force security elements land via assault support.
- Dispersion of Assets. The required degree of dispersion is reflected in the design of the LOA and planned array of forces within and EAB. Preparation and implementation of the littoral maneuver plan must account for the sequence of critical capabilities required to achieve desired effects.

Task Organization of Littoral Maneuver Squadron (LMS). As discussed in section 6.4.3, the LMS is the task-organized collection of LSMs and other vessels that enable movement of the littoral force to and within LOA(s). The type and availability of vessels and craft composing the LMS is informed by the concept of operations. At times, elements of the LMS supporting a particular EAB may support operational-level logistics while also supporting tactical-level maneuver. The littoral force commander and commander of the LMS must coordinate to ensure accomplishment of the SOM in conjunction with sustainment logistics activities.

Organization for Embarkation. The organization for embarkation must support both the plan for littoral maneuver and the SOM. It must also provide for maximum flexibility to support multiple loading sites both ashore and at sea. The tenets of combat loading guide the arrangement of personnel and stowage of equipment to facilitate the anticipated tactical operation.

Control of Seaward and Landward Operations. There are multiple methods to control seaward and landward operations. Possible methods range from the possible adaptation of current amphibious doctrine to a tailored method based on the envisioned littoral force relationship with other forces afloat. In all of these relationships, command authorities and responsibilities, planning considerations, and command relationships have a commonality. The forces afloat controlling seaward operations must fully integrate with fleet operations and maneuver at sea. EABO experimentation will require testing of the littoral force commander's ability to integrate forces from the seaward side and transition to a landward operation. These transitions may occur continuously throughout the operation.

7.4.4 Force Protection and Expeditionary Advanced Base Security

Force protection is the ability of the littoral force to avoid or withstand adversary actions or environmental conditions while retaining the ability to fulfill its primary mission. SIGMAN, camouflage, concealment, and deception operations enable the littoral force to avoid or mitigate the effects of adversary actions. During campaigning, state, non-state and proxy actors and networks can help or hinder security. Adversary SOF forces, hiding among the local population, can gain information on EAB location, disposition, and strength. Civil public demonstrations and violent crowds can also affect EAB security, requiring the need and use of intermediate force capabilities.

Security is critical at and within the various sites that compose an EAB. During crisis to conflict, adversary attacks may take the form of air and missile strikes, naval gunfire, special-forces raids, attacks in the EM spectrum, and activities within the IE. The littoral force will combine active and passive measures to ensure the preservation of capabilities and overall combat power.

Local Site Security. Each task-organized element of the littoral force possesses selected capabilities to accomplish the SOM. The LFC assesses baseline activities and the threat level in the locality where a given element is operating. The composition and size of the task-organized element will be tailored to each EAB based on the assessed threat level, assets requiring protection, and proximity to adjacent units. The LFC must weigh the trade-offs between increased security-element size and added signature, movement requirements, and sustainment needs.

7.4.5 Signature Management

The littoral force may employ SIGMAN techniques to disrupt target acquisition, tracking, and terminal guidance. These techniques include both offensive and defensive actions. In keeping with the characteristics of EABO, they are primarily passive measures and must be considered by all elements of the littoral force. In terms of force protection, SIGMAN is an element of OPSEC, and the littoral force must manage signatures for two fundamental reasons:
- Support force protection (Survivability)
- Achieve surprise (Deception)

This section focuses on requirements and planning considerations necessary to perform SIGMAN in support of littoral-force survivability. Although many considerations are the same, SIGMAN in support of deception is discussed in chapter 3, "Intelligence Operations," and chapter 4, "Operations in the Information Environment."

SIGMAN Requirements. SIGMAN depends on knowledge of *indicators*, which are detectable, friendly actions and open-source information that an adversary can interpret and piece together to derive critical information about the littoral force. Indicators have one or more of the following characteristics:
- Signatures, which are observable activities and operational trends that reveal critical information to adversary intelligence collection

- Profiles, which refer to the sum of unique signatures and associations generated by a functional activity
- Associations, which make an indicator identifiable or cause it to stand out
- Contrasts, which refer to the differences observed between an activity's standard profile and its most recent or current actions
- Exposures, which refer to when and for how long an indicator is observed

Indicators may also be categorized as physical, technical or administrative.[46] *Physical signatures* are those the adversary can collect by direct observation or geospatial-intelligence assets. Detection of *technical signatures* typically requires specialized equipment, such as adversary SIGINT or measurement and signature intelligence (MASINT) assets. Individuals and units create *administrative signatures* when planning, maneuvering, contracting for support, and performing other administrative actions. An adversary observes or detects administrative signatures via human intelligence (HUMINT), SIGINT, OSINT, and OCO.

Understanding how an enemy perceives friendly indicators is the crucial second step, informed by intelligence processes of conduit and kill-chain analysis described in chapter 5, "Intelligence Operations." Insights gained from these processes lead to the third step of SIGMAN: countermeasures taken to reduce, modify, or display indicators to achieve commander's intent.

Own-Force Signature Assessment (OFSA). The littoral force must continually integrate the OFSA process with training and exercises to refine and incorporate SIGMAN procedures into operations. The ability to understand own-force signatures supports a functional SIGMAN process, and it is a critical requirement to enable assessment of countermeasures effectiveness or determine future signature protection capabilities. The littoral force must include OPSEC considerations in the planning process and assess indicators during operations. OFSA validates baseline data in order to capture unintended or unexpected indicators using a variety of collection methods.

OFSA requirements must be balanced with the tempo and scale of operations. Increased use of overt OFSA may be appropriate for exercises and experimental TTP development, whereas during combat operations OFSA may be limited to EMCON monitoring of high-value C4 nodes, actions by individuals or small unit leaders, or incidental collection. Self-awareness coupled with commander's intent then allows Marines at all levels of leadership to manage select indicators and manipulate their exposed profiles in relation to the adversary capability.

Electromagnetic Signature Control. Some techniques for countering electromagnetic signature vulnerabilities lie within the realm of sophisticated technology such as highly directional or low-probability-of-intercept/detect transmissions. However, many effective countermeasures to increase electromagnetic protection are derived from traditional OPSEC disciplines employed for electromagnetic signature control. These may include:
- Setting EMCON conditions that restrict use of own force electronic systems to specific periods or conditions
- Conducting DCO
- Remotely locating transmitters and/or antennas at some distance from a C2 or communications nodes
- Using highly directional antennas to reduce the signal strength available for adversary exploitation

---

[46] Joint Chiefs of Staff, *Operations Security*, JP 3-13.3 (Washington, DC: US Department of Defense, 2016) and Joint Chiefs of Staff, *Military Deception*, JP 3-13.4 (Washington, DC: US Department of Defense, 2017).

- Using wire and fiber optics pathways for communications
- Using frequency hopping/spread-spectrum radios
- Employing a SOM or avenues of approach that interpose terrain between friendly transmitters and adversary intercept stations
- Employing sound radio discipline, to include using minimum required power
- Using brevity codes and digital communication
- Reducing reporting requirements and unnecessary traffic (e.g., routine communications checks)
- Employing proper terrain-screening crest by radars
- Operating effectively with a smaller technical footprint and the decreased information that results

Intelligence in Support of SIGMAN. For SIGMAN intelligence requirements, see the discussion in chapter 4, "Intelligence Operations."

Counter-Reconnaissance. Counter-reconnaissance activities degrade or deny an adversary's ability to determine composition, disposition, and strength of an EAB and other friendly forces. Baseline intelligence collections and reconnaissance activities set the initial conditions for the counter-reconnaissance effort. Counter-reconnaissance activities may be either passive or active depending on the assessed threat level and requirement. Counter-reconnaissance activities may include:
- Local security patrols
- Hide site employment
- Information collection from the local populace
- Unmanned aerial system operations
- Fixed-wing, rotary-wing, or tiltrotor aviation operations
- Visual deception
- Electromagnetic spoofing

Counter-reconnaissance efforts also look across all other domains to sense anomalies in the baseline and contribute to commander's situational awareness and ability to detect adversary efforts to disrupt littoral force activities.

Passive Defensive Measures. Camouflage and concealment use materials and techniques to hide, blend, disguise, or disrupt the appearance of military targets and/or their backgrounds. Effective use of these materials and techniques degrades the effectiveness of adversary intelligence, surveillance, reconnaissance, and targeting capabilities. Obscuring or altering telltale signatures of units on the battlefield can defeat both skilled observers and sophisticated sensors.

To be effective, camouflage, concealment, and signature alteration must function in the frequency wavebands that will degrade sensor and seeker performance sufficiently to deny or delay targeting and weapon guidance. Consequently, a variety of systems and techniques are necessary to defeat the range of possible adversary battlefield sensors and munitions seekers. Camouflage, concealment, and signature alteration actions are an essential part of tactical operations. Littoral forces must practice them with discipline and continuously integrate them into mission planning and IPB processes.

Movement and Dispersion. Movement supports denial of adversary targeting by repositioning friendly forces at a rate faster than the adversary network can effectively complete its decision cycle. Dispersion stresses the adversary's kill chain by greatly increasing the number of potential targets while also reducing the potential lethality of an individual round. Greater dispersion reduces the efficiency of adversary target acquisition, since more ISR assets are required to find the same number of targets. As countermeasures, effective movement and dispersion require both careful planning and a clear understanding of the adversary decision cycle. Too frequent movement, however, may increase opportunities for adversary collection.

7.4.6 Defensive Fires Plan

Air and Missile Defense Fires. The littoral force, in concert with the RADC, SADC, and/or AMDC, conducts AMD operations to protect vital areas and defend critical assets. Defensive fires in the air domain involve the fundamental practice of constructing a kill chain. Key elements of AMD include:
- Determining critical assets and vital areas
- Applying AMD assets to critical assets to generate a defended asset list
- Granting authorities to air-control agencies to manage the defense of the vital area
- Employing missile engagement zones, fighter engagement zones, and joint engagement zones by the controlling agency
- Exchanging timely information between air- and missile-defense assets (including sensors, firing units, and C2 agencies)
- Employing tactical data links and voice communications
- Planning missile defense with the wider naval and joint forces

Refer to *Countering Air and Missile Threats*, JP 3-01, for detailed joint doctrine concerning AMD and *Composite Warfare: Maritime Operations at the Tactical Level of War*, NWP 3-56, for comparable naval doctrine.

Other Fires Planning. While the littoral force provides fires to support assigned missions and objectives, it must also plan for the employment of fires to preserve combat power. The littoral force may establish boundaries or control measures between echelons possessing fire support coordination centers (FSCCs) to facilitate the responsive employment of fires. The LFC will assess threat levels and determine the size of security elements needed to preserve the combat power of fires capabilities.

External Agencies. To leverage external fire-support agencies in defense of the littoral force, the senior FSCC within the LOA must maintain sufficient battlespace awareness to approve fires into the LOA and potentially within close proximity of friendly forces.

7.5 MISSION CONCEPTS OF EMPLOYMENT

The following subsections offer concepts of employment (CONEMPs) for use by forces conducting EABO missions and tasks. These CONEMPs reflect the developing capabilities associated with future force design. They are presented by system and/or littoral force element to illustrate in isolation the key requirements and employment considerations for each. When combined, these systems employed by task-organized littoral force elements create the units of action that comprise an EAB. Figure 7-1 below on page 7-9 illustrates combined actions by littoral forces conducting EABO.

7.5.1 Fires in Support of Surface Warfare

The littoral force plays a vital role within the greater naval force by applying fires against maritime surface targets to deny or control sea space (see figure 7-2 below on page 7-10). Fires in the maritime domain fundamentally require construction of a maritime fires kill chain that results an effective, efficient ability to hold adversary forces at risk and protect vital areas and units. These fires consist of land-based, sea-based, and air-launched missiles and loitering munitions delivered from manned and unmanned

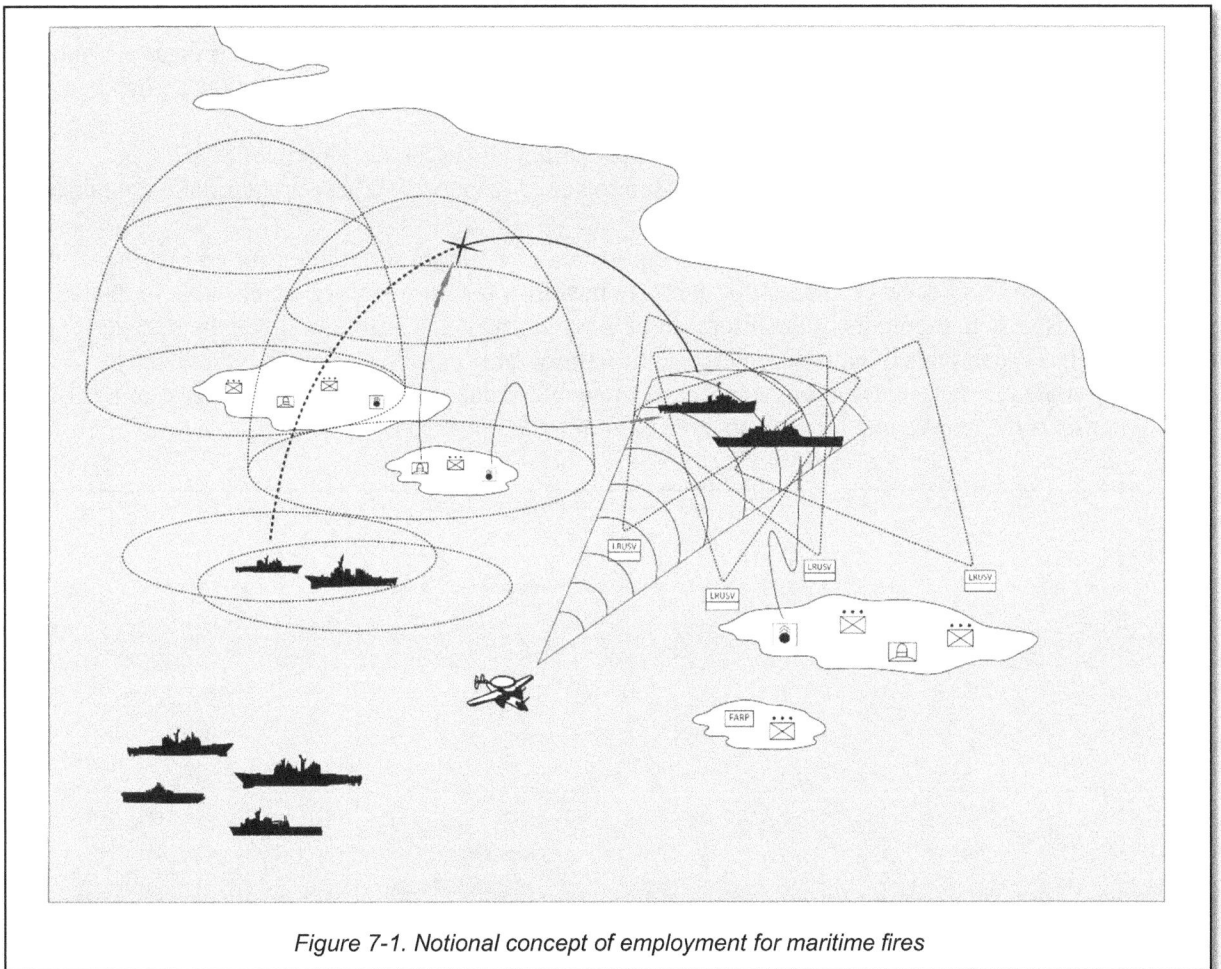

Figure 7-1. Notional concept of employment for maritime fires

platforms. The littoral force may control fires using the methodologies of either the MAGTF fires framework or composite warfare construct. Mission-engineering threads will vary based on the C2 methodology in effect and assets employed to accomplish the SOM. Key requirements follow:

- Network. The littoral force's firing units must be capable of receiving firing data from multiple sources: forward observers, reconnaissance assets, aircraft, adjacent units, tactical headquarters, or even directly from the MOC. Regardless of the source, firing units receive targeting data directly rather than through several echelons of the task organization. Receipt of firing data in this direct fashion implies precoordination of the mission, airspace deconfliction, and approval by the very nature of the transmission. Organic to the littoral force, the commander needs sufficient communications capabilities to mass fires in time and space from both land- and sea-based manned and unmanned platforms that vary significantly in speed, trajectory, munitions effectiveness, and quantity.
- Sensing. Sensing requirements will vary in complexity. Within composite warfare, the littoral force must maintain communications with the SUWC or SCC and contribute sensing capabilities to the combined force. However, the littoral forces may not require the same level of maritime domain awareness as the SUWC/SCC. Situations where the appropriate tactical commander directly provides targeting information to a firing unit reduce requirements for maritime domain awareness.
- Classification and Identification of Maritime Objects. The littoral force must have the ability to distinguish threats from benign objects among the maritime clutter or receive this clarity from another source.

- Decision Making. The littoral forces must have appropriate authority and capability to make a timely, correct, and lawful determination to engage the threat; determine the appropriate firing unit and weapons system; issue a firing command; observe the effects; and conduct a damage assessment.
- Preplanned Responses. The littoral force must participate in the development of and be thoroughly familiar with the naval force's preplanned responses. These are normally contained in the classified OPTASK.
- Site Selection. The firing battery must be positioned on terrain suitable for the employment of its munitions, have access to numerous alternate positions for survivability, and be able to maneuver and disperse to the limits of the littoral force's sustainment capabilities. Given the missile defense capabilities of adversary vessels, the battery must be able to mass geographically disparate platoons and sections as necessary to achieve desired effects. Commanders must also consider the intervening terrain and overflight of noncombatants.

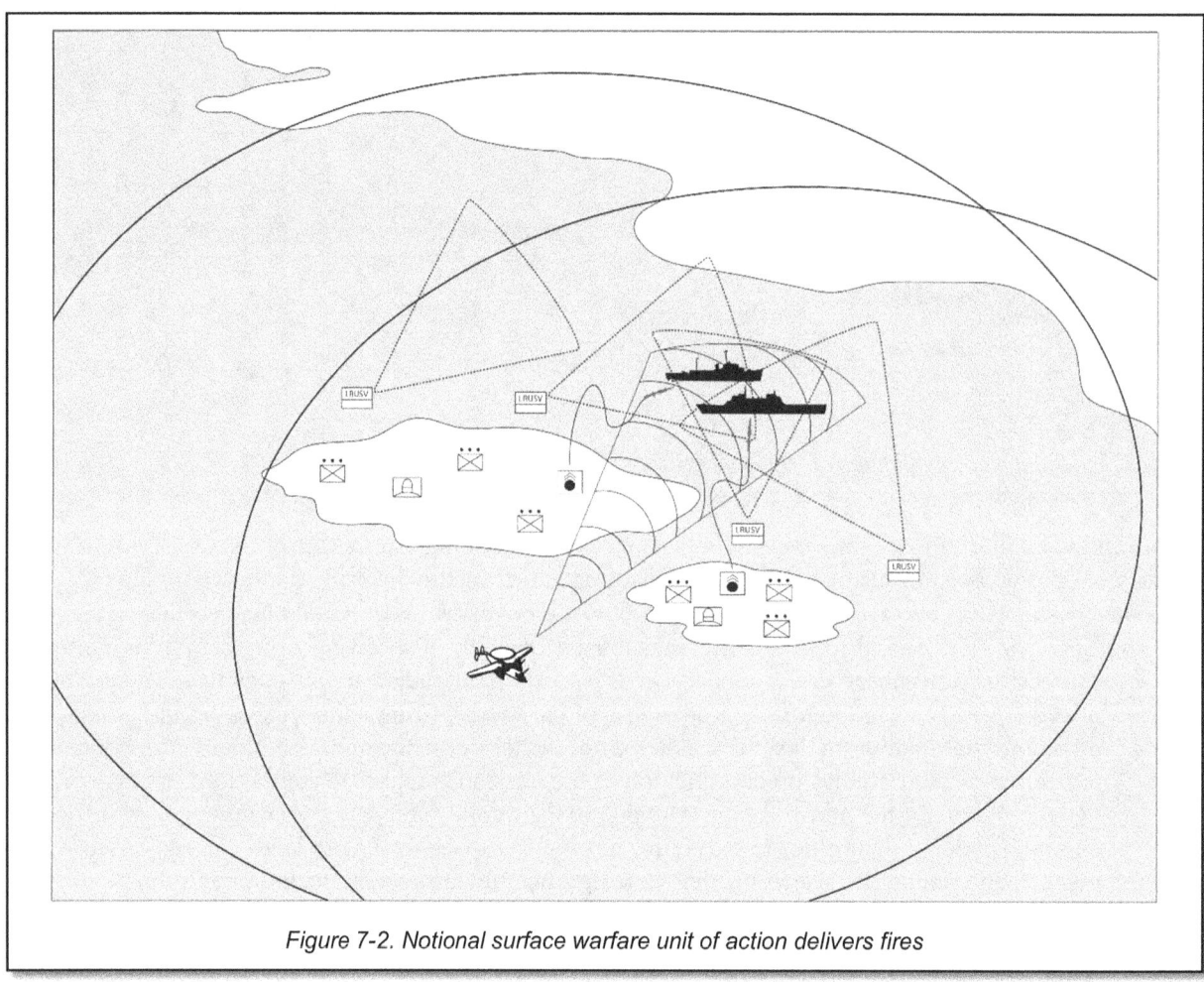

Figure 7-2. Notional surface warfare unit of action delivers fires

Employment considerations follow:
- Units equipped with an optionally manned, long-range unmanned surface vessel (LRUSV) may be organized into platoons, but their employment need not reflect this organization. LRUSV units require coastal terrain sufficient to mitigate adversary threats from land and sea, as well as sanctuary to protect against adverse weather effects. LRUSV platoons must consider a tactical rotation aligned with these considerations.

- LRUSV platoons must assess and account for unique requirements associated with maintenance, rearming, and refueling autonomous systems.
- LRUSV platoons should consider one-for-one vessel rotations to support maintenance, rearming, and refueling.
- While LRUSVs combine sensing and shooting capabilities, the commander must consider a balanced employment of these capabilities. Forward positioning facilitates timely employment of organic precision fires but exposes the vessel to adversary engagement. The commander must weigh the value of the LRUSV as a reconnaissance asset versus a fires platform.

7.5.2 Fires in Support of Air and Missile Defense

The littoral force conducts defensive actions to destroy, nullify, or reduce the effectiveness of hostile air and missile threats against friendly forces and assets (see figure 7-3 below). AMD will primarily be conducted using joint and organic short-, medium-, and long-range surface-to-air weapons with indications and warnings provided by external sources along with local, passive sensors. Engagements may also use active sensors. Key requirements follow:
- For AMD, the littoral force must be capable of integrating with an external network to receive or organically sense joint tracks.
- The littoral force must be capable of command and control in the naval and joint systems to employ the common operational picture, tactical networks, and data systems in the execution of preplanned responses.

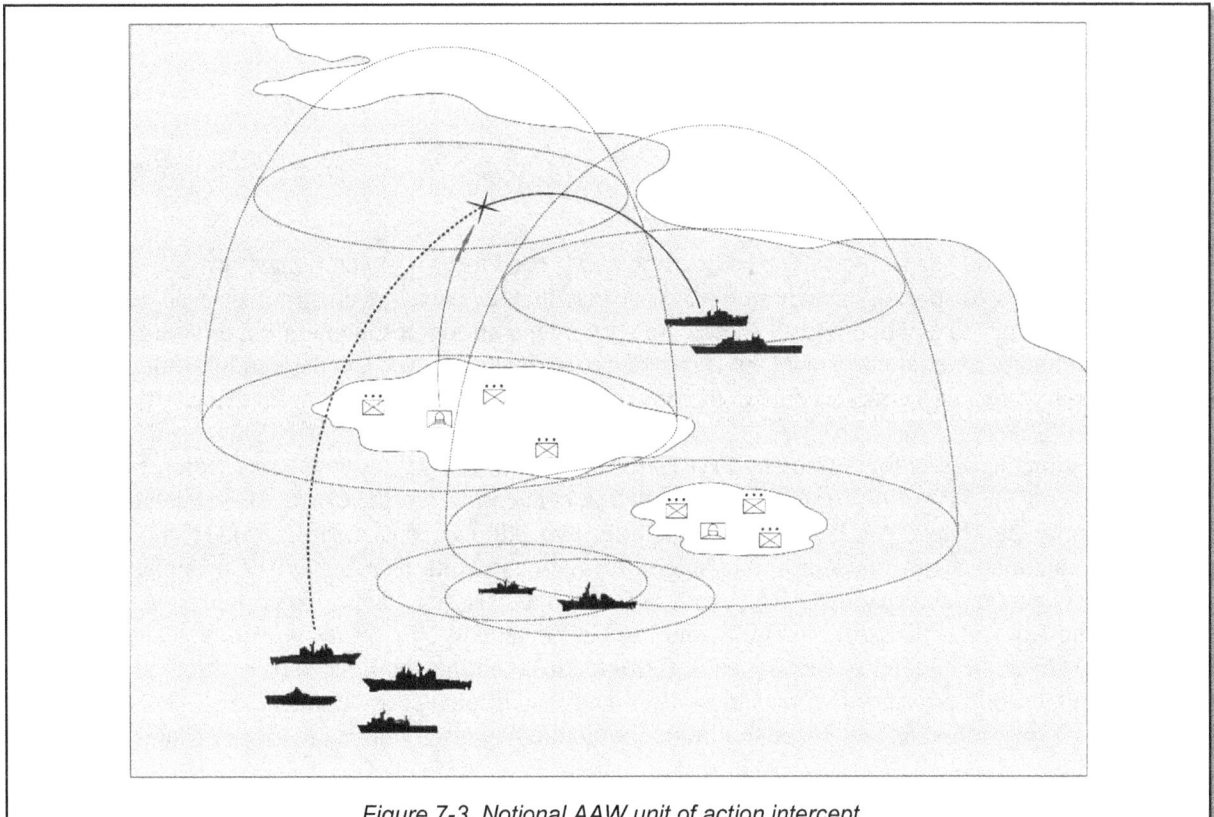

Figure 7-3. Notional AAW unit of action intercept

Employment considerations follow:
- The commander must understand the requirements for continuous or varying AMD coverage. Based on threat analysis, coverage requirements may vary between persistent or episodic.
- Given the range of AMD capabilities, commanders must carefully position assets requiring protection, as well as the vital area specified by the warfare commander. Positioning these capabilities may require balancing these potentially competing requirements.
- Considering the relative difficulty of sustaining various elements of the combined force, the commander must be prepared to recommend the appropriate employment of AMD weapons in order to mass effects. Ultimately, the commander must ensure the efficient employment of available munitions to conserve finite resources.
- Similar to surface warfare capabilities, AMD assets must displace frequently for survivability. The littoral force must consider displacing by echelon in order to mitigate capabilities gaps while providing protection to the force and within the vital area.
- AMD units will likely be employed as elements of a larger task-organized group afloat or ashore.

7.5.3 Operations in Support of Antisubmarine Warfare

The littoral commander may task organize an EAB or multiple EABs to support ASW and be responsive to the ASWC when operating under composite warfare. The employment of ASW capabilities will enhance the scouting and anti-scouting of sustained theater-level undersea warfare campaigns. EABs operating in support of ASW provide sensing and data collection for the maritime COP while also enabling forward logistics and support. The littoral commander may also have the ability to emplace defensive and protective sea mines after tasking or coordination with the mine warfare commander.

7.5.4 Support to Information Warfare

For IWC planning considerations and employment, refer to sections 4.5 through 4.8.

7.5.5 Forward Arming and Refueling Points

Marine Corps aviation has well-established doctrine and procedures for expeditionary airfield and FARP operations, which are especially salient in EABO because they increase the operational reach of aviation forces, add resilience to aviation logistics, increase sortie-generation rates in operating areas, and increase flexibility in the use of aviation. Given the flexibility envisioned in EABO, consideration should also be given to employing multiple replenishment methods in support of platforms other than aircraft.

The MWSS contains assets for employing FARPs that support Marine Corps aviation. Refer to *Aviation Ground Support*, MCTP 3-20B, for detailed information concerning Marine Corps AGS in general and FARPs in particular. Naval Air Training and Operating Procedures and Standardization (NATOPS) manuals contain mandatory procedures to enable safe aviation operations in austere environments and preserve sortie generation capability. Relevant NATOPS manuals include the Expeditionary Airfield NATOPS, the Air Traffic Control NATOPS, and NATOPS manuals by aircraft type, model, and series to use FARPs. Given the flexibility envisioned in EABO, consideration should also be given to employing FARPs in support of platforms other than aircraft. Finally, additional considerations for FARP operations include HN coordination and adherence to minimum manning levels contained in technical and safety directives.

Key requirements and planning considerations follow:
- Refueling Operations. Operational requirements, including duration of sustainment and type of equipment being serviced, influences the decision on the type of aviation refueling mission to be executed (i.e. hot or cold FARP, flight line refueling, etc.). This choice will affect the

responsiveness and rapidity of aircraft employment, but planners must evaluate the benefits of a hot FARP against limitations such as quantity of aircraft refueling, crew rest, and planning time for further missions. If employing the FARP to support platforms other than aircraft (i.e. ground equipment, unmanned systems, or maritime vessels and craft), consideration must be given to the different types of required fuel (especially if using HN/contracted equipment), pumps, storage containers, storage requirements, distribution piping and nozzles as well as the personnel to operate these different systems.
- Ordnance. In an intensive naval campaign, particularly as competition escalates, aircraft may require ordnance before they run out of fuel. It will likely be necessary to pre-position ordnance in forward areas and to resupply FARPs from ports and airfields. Requirements for ordnance crew, especially for handling specific munitions and Type/Model/Series aircraft, will determine the number and type of personnel to support the FARP.
- Expeditionary Airfield (EAF). Tasks associated with establishing and operating an EAF include the layout and setup of the site, to include the proper spacing, marking, and lighting for aircraft operations.
- Expeditionary Firefighting and Rescue (EFR). For FARP operations during contingency operations, EFR support is not required. However, safety concerns dictate that, when available, one EFR apparatus and rescue vehicle should support a FARP.
- Controlling Agencies. Requirements for controlling agencies such as a Marine ATC Mobile Team are based on the type, model, and series of aircraft that the FARP is supporting, volume of aircraft expected at the FARP, requirement for expedient landing-zone survey, and airspace management considerations.
- Force Protection. Airborne aircraft in the vicinity of a FARP and grounded aircraft are susceptible to both air and ground fires. Consideration must be made for AMD and ground security to protect sortie generation.

7.5.5.1 Forward Arming and Refueling Point Planning

During EABO, FARPs must operate in small, highly mobile units able to rapidly deploy and effectively manage their signatures. In moderate- or high-threat environments, FARPs must move frequently to avoid detection. In a low-threat environment with a static front and little enemy air activity, FARPs may displace less often. Varying situations will determine whether to employ multiple FARPs or rapidly relocate a single FARP between multiple sites.

During establishment of multiple FARP sites or relocation of a single FARP, the new FARP should be operational before the operating FARP is shut down. Speed of movement to establish the FARP site is of prime importance, and planners must allocate adequate time to set up equipment. Three primary FARP methods, features of which may be combined in execution, exist for employment:
- Ground Transported. Establishing a FARP using ground vehicles is the most common means of employment.
- Air Delivered. Using assault support assets is an alternate means of establishing a FARP. Assault support aviation assets can also directly pass fuel to other aircraft or ground vehicles. Operators employ air-delivered FARPs in tactical operations requiring rapid emplacement or when ground transportation is infeasible due to insufficient assets or inhospitable terrain.
- Surface Transported. The operational situation may dictate establishing a FARP using surface connectors. Once ashore, the FARP operates in the same manner as a ground-transported FARP. These FARPs are logistically flexible and do not require use of aviation assets for setup or resupply. Surface-transported FARPs are preferred when the tactical situation, terrain, and time do not allow for the movement of ground assets into the desired location.

Important considerations in selecting the FARP method include:

- Coordination requirements and potential logistic sustainment
- Availability and type of surface connectors (both manned and unmanned), aircraft, or landing craft
- Location of CSS areas
- Landing locations and obstructions
- Main supply routes
- Distance to the FARP
- Timing
- Threats and security requirements

7.5.5.2 Forward Arming and Refueling Point Designs

Factors such as mission requirements, aircraft constraints, such as weight and maneuverability, and environmental constraints will determine the type of FARP necessary to enable operations. The LFC and staff must understand this because each type of FARP comes with varying requirements relating to size, personnel, equipment, engineering preparation, and sustainment. Marine Corps FARPs do not currently support all US Navy aircraft. AGS planners must conduct detailed planning in coordination with counterparts from the Navy Expeditionary Combat Command, supported flying squadrons, and subject matter experts in Navy-specific aircraft.

FARP designs are contingent on the type of aircraft being supported. For EABO, there are two primary support layouts detailed in Appendix D of *Aviation Ground Support* MCTP 3-20B: (1) an *assault FARP* for rotary-wing and tiltrotor aircraft without forward-firing ordnance and (2) an *attack FARP* for rotary- and fixed-wing aircraft with forward-firing ordnance. For assault FARPs supporting fixed-wing aircraft, the ability to taxi from arm/de-arm headings is required. Ordnance operations can take place in an assault FARP based on the type, model, and series of aircraft supported, and qualified aviation ordnance and aviation maintenance personnel. Assault FARPs normally require a larger area for operations due to the required safety distances.

7.6 FLEET INTEROPERABILITY

As stated in chapter 1, EABO take place within the context of a larger naval campaign. As such, EABs are designed to complement the seagoing elements of a fleet. On occasion CSGs, ESGs, SAGs, ARG/MEUs, and combat logistics forces execute tasks to enable or support EABO as part of a littoral force. For future experimentation, the following are potential missions:
- Support movement of littoral forces into the fleet AO and LOA
- Support seizure of maritime terrain
- Support establishment of an EAB
- Support defense of an EAB
- Support sustainment of littoral forces

APPENDIX A

FUTURE FORCE DESIGN AND CONSIDERATIONS

A.1 GENERAL

This appendix only contains force design considerations for future MLR and the infantry battalions that make up these units. III MEF continues efforts to stand up the first MLRs, and I MEF is working to demonstrate EABO capabilities with current organization and equipment. While other properly task organized elements of current MAGTFs may be able to contribute to EABO and execute sea denial operations, they represent modifications to current forces and are thus part of force development vice force design.

A.2 2030 MARINE LITTORAL REGIMENT (MLR)

The 2030 MLR will maneuver and persist inside a contested maritime environment and conduct sea-denial operations as part of the naval expeditionary force to enable fleet operations. It must be designed to:
- Conduct surveillance and reconnaissance
- Generate, preserve, deny or project information
- Conduct screen/guard/cover
- Deny or control key maritime terrain
- Conduct surface warfare operations
- Conduct air and missile defense
- Conduct strike operations
- Conduct sustainment operations

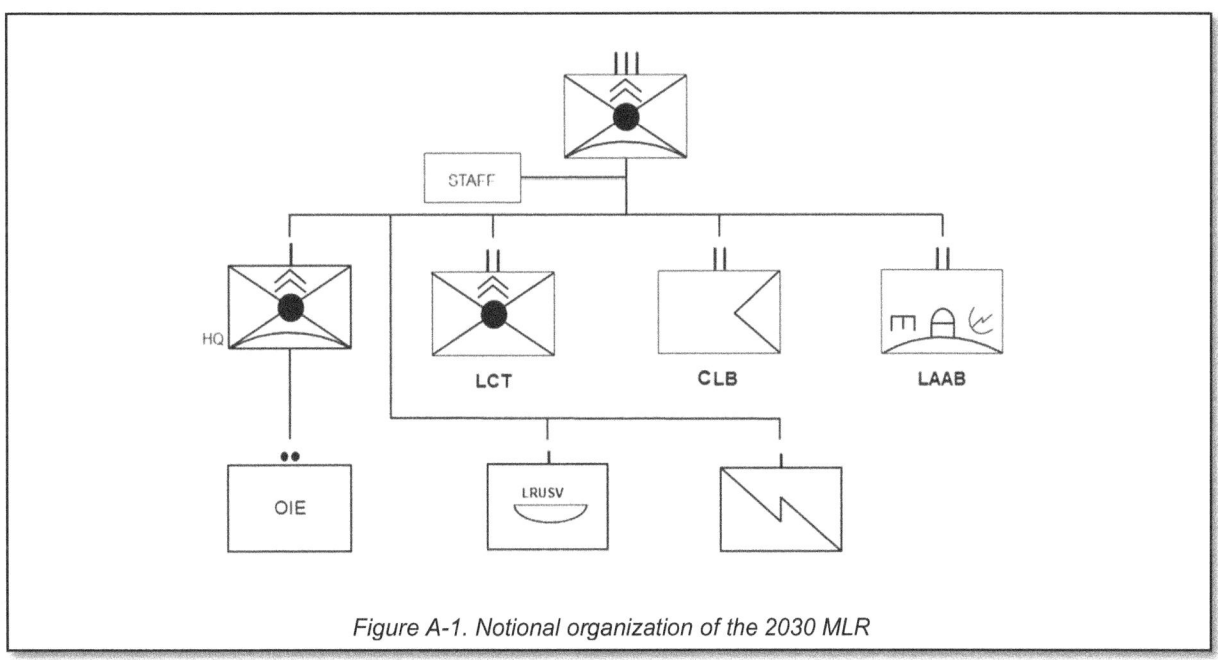

Figure A-1. Notional organization of the 2030 MLR

The MLR will be composed of a headquarters with fires (lethal and nonlethal), a littoral combat team, combat logistics battalion, and littoral antiair battalion.

A.2.1 Littoral Combat Team (LCT)

The LCT will be employed as a task-organized maritime littoral unit, capable of commanding and controlling distributed EABs that are conducting sustained operations to enable fleet operations via sea denial. The proposed 2030 infantry formations from team to battalion will generate infantry units to support various missions via infantry core mission-essential tasks (METs). These formations will provide the core forces for battalion landing teams (BLTs) and LCTs, while preserving infantry battalions proficient in core METs for conventional major combat operations.

The LCT will be formed on the foundation of an infantry battalion with attachments. It will be task-organized to provide an MLR commander the capability to conduct a variety of operations, including fires and/or reconnaissance/counter-reconnaissance, across multiple EABs, all operating under the C2 of the LCT headquarters.

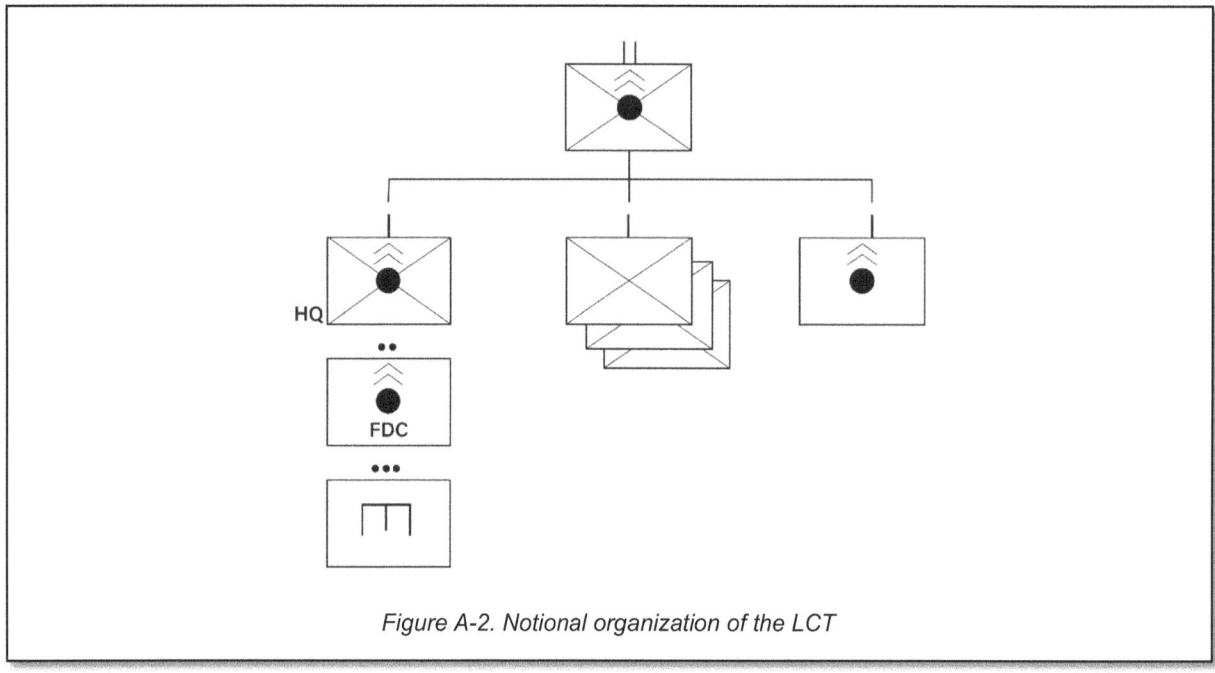

Figure A-2. Notional organization of the LCT

A.2.2 Combat Logistics Battalion (CLB)

The MLR CLB will provide tactical logistics and EOD support to sustain regimental operations across the competition continuum. The CLB's core tasks include the following operations and services:
- Prepositioning
- Ground Supply
- Ground equipment maintenance
- Littoral transportation
- General engineering
- Health services
- Operational contracting
- Explosive ordnance disposal

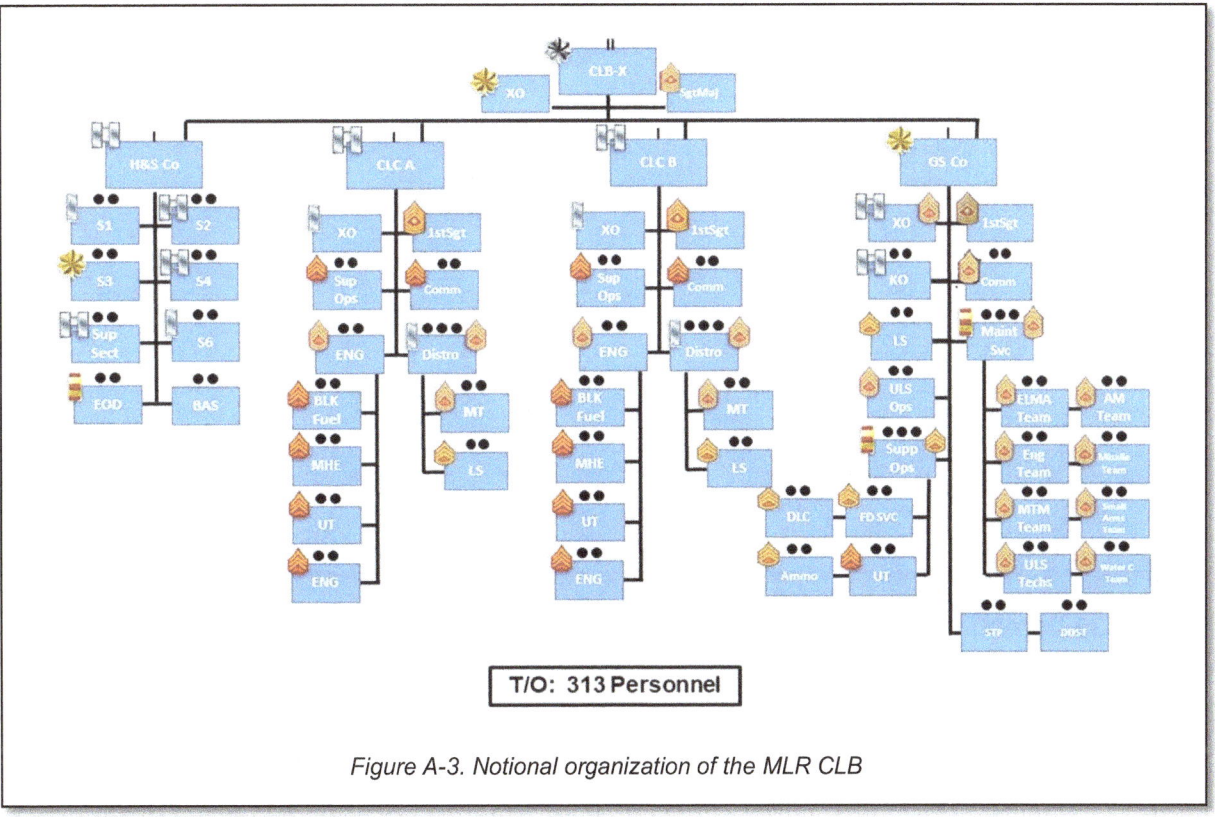

Figure A-3. Notional organization of the MLR CLB

A.2.3 Littoral Antiair Battalion (LAAB)

Sourced from the MAW, the LAAB will be a composite battalion that includes elements from the MWSS, Marine wing communications squadron, Marine air support squadron, Marine air control squadron, and ground-based air defense.

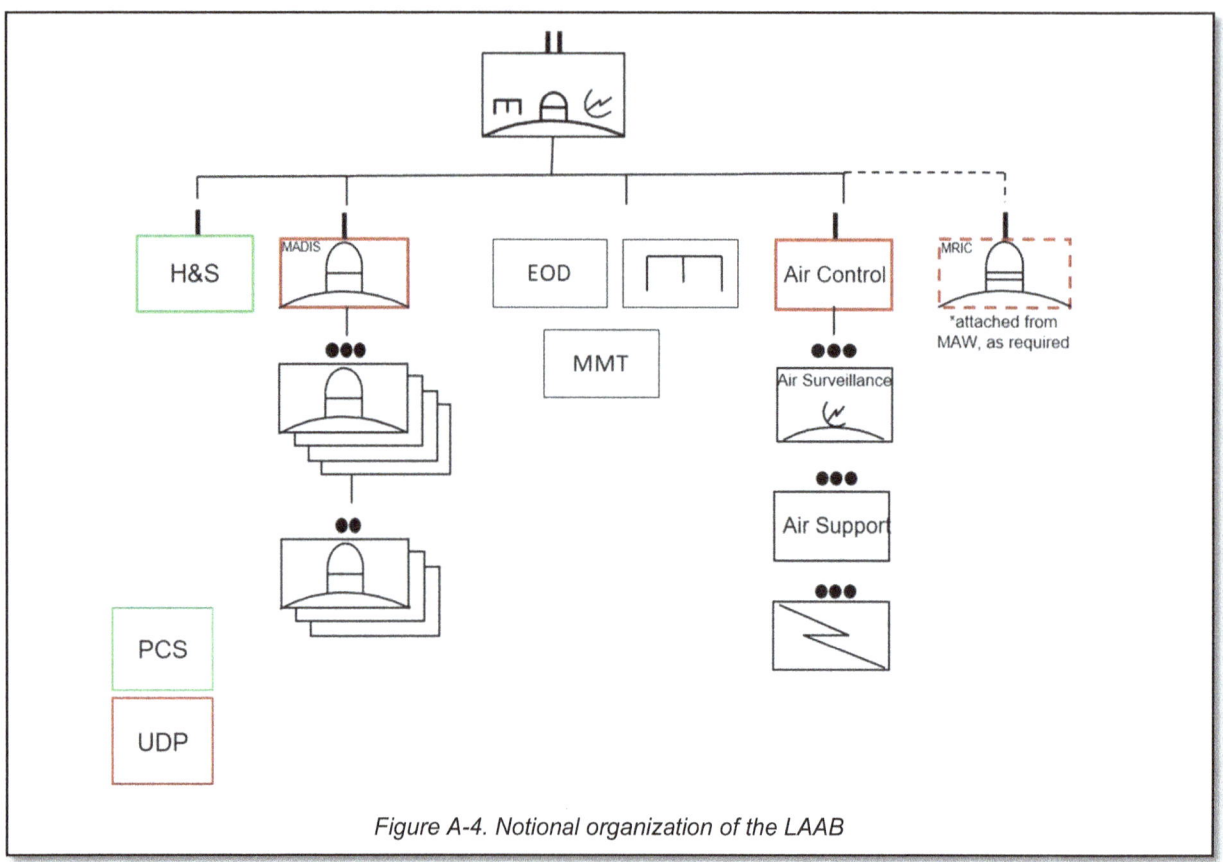

Figure A-4. Notional organization of the LAAB

A.3 2030 INFANTRY BATTALION

The 2030 Marine infantry battalion will contribute to joint and naval combined-arms formations that are essential components of the future, persistently forward-deployed naval expeditionary force. Optimized to operate in the contact layer, these transformed infantry forces will execute mission-critical tasks for the fleet commander or maritime component commander, often in conjunction with or in support of SOF partners. To accomplish their tasks, infantry battalions must be organically equipped, starting at the squad level, with resilient, networked communications and precision fires capabilities, including loitering munitions enabled by artificial intelligence. These units must be light, mobile, and capable of distributed operations. They must be able to embark aboard all types of Navy and auxiliary vessels. And they must be armed with organic systems capable of sensing, cueing, and shooting in support of naval and joint sea-control and assured-access missions. Mature, competent, highly trained and educated Marines equipped with state-of-the-art weapons and equipment are essential to achieving this vision.

Infantry battalions will deploy rotationally as the LCT of an MLR or as the BLT of a MEU. Rotational deployments to the Western Pacific maintain persistent forward presence to support allies and partners, deter aggression, and if attacked to fight and win. They will also deploy in response to crisis as part of an infantry regiment or special purpose MAGTF. Core mission-essential tasks for the 2030 infantry battalion will include:
- Conduct expeditionary operations

- Conduct offensive operations
- Conduct defensive operation
- Conduct crisis response

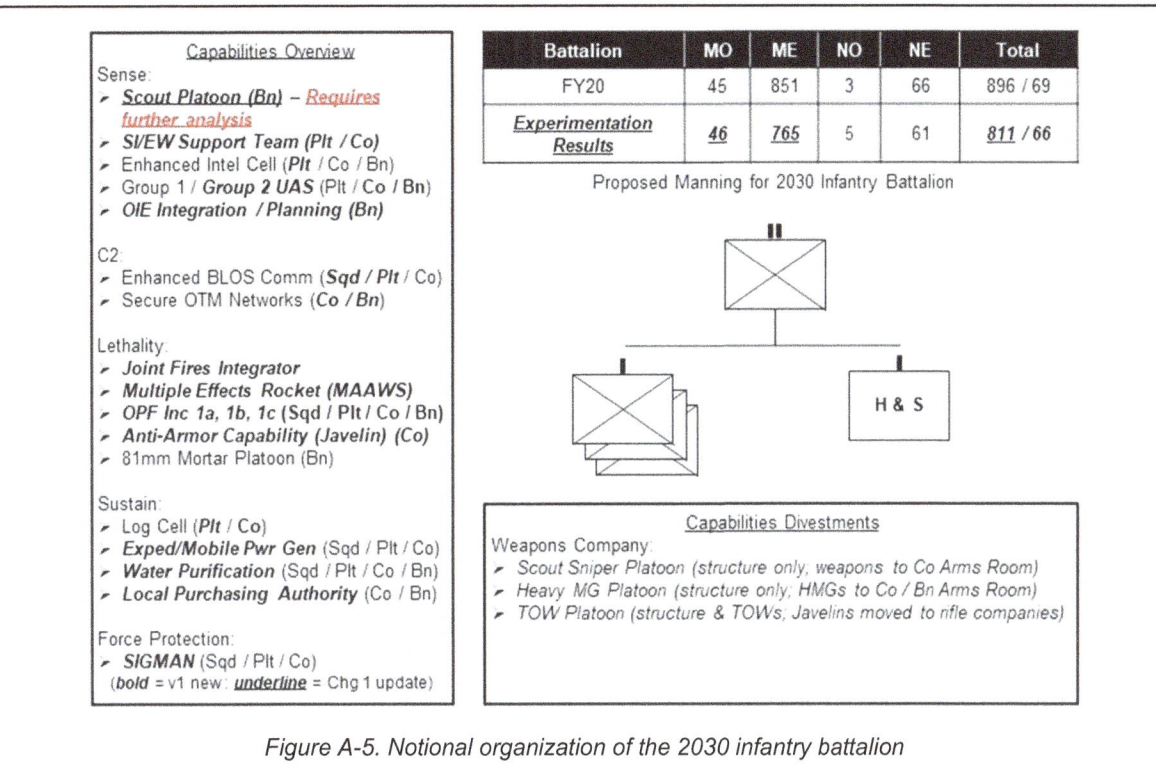

Figure A-5. Notional organization of the 2030 infantry battalion

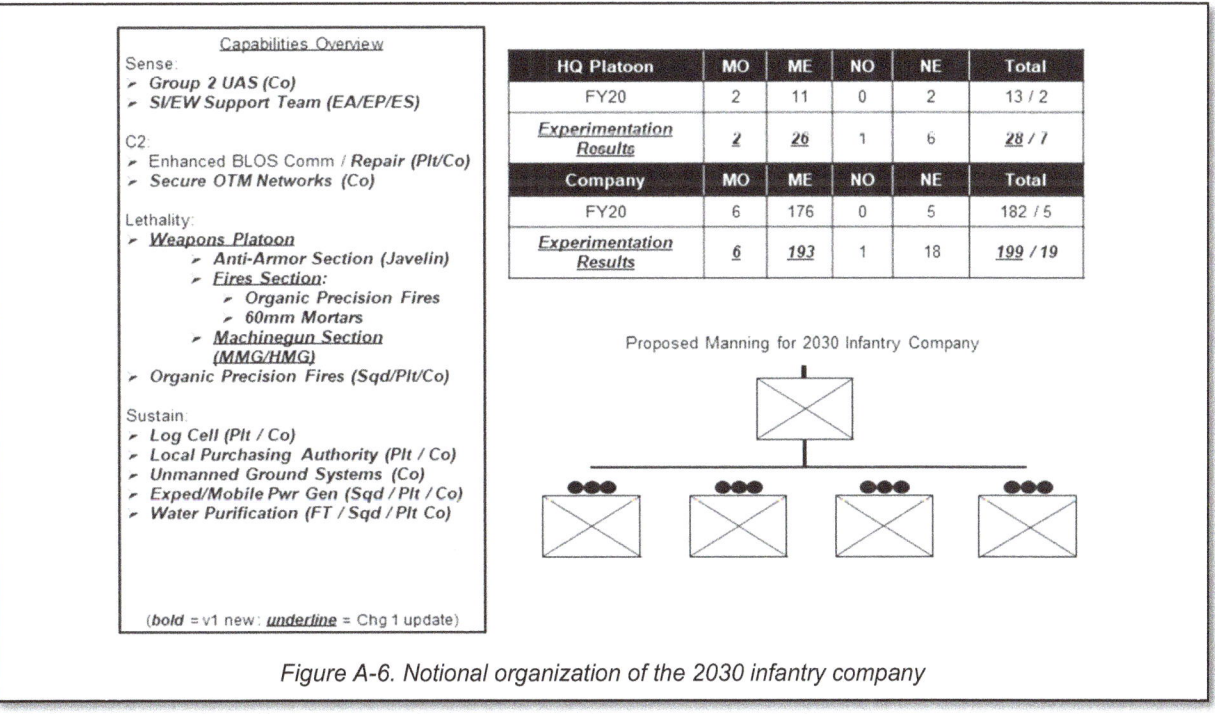

Figure A-6. Notional organization of the 2030 infantry company

INTENTIONALLY BLANK

APPENDIX B

MISSION-ESSENTIAL TASKS

B.1 GENERAL

Deputy Commandant for Combat Development and Integration (specifically Marine Corps Task List (MCTL) Branch, Marine Corps Integration Division, Capabilities Development Directorate) has refined and reviewed EABO Marine Corps Tasks (MCTs) for the MLR and MEU following the 1st edition of TM EABO. The review included a matured organizing, training, and equipping requirement framework, understanding the associated qualitative and quantitative standards when they are chosen as METs and included in a mission-essential task list (METL). Final adjudication of the MLR METL was completed in July 2022, which resulted in the adoption of 16 new MCTs that have since been incorporated into the MCTL. All METLs are DRAFT and stand to serve as the foundation for readiness reporting in the Defense Readiness Reporting System once validated and approved. The MCTs listed below will undergo continuous, rigorous review and revision over the next few years based on findings from experiments, war-games, and service-wide exercises.

Marine Corps task revisions should codify EABO for core and assigned mission essential tasks across the FMF as applicable. This will continue to inculcate EABO into the Marine Corps as a whole rather than resident within a few select units.

B.2 PROPOSED MARINE LITTORAL REGIMENT MISSION-ESSENTIAL TASK LIST

B.2.1 Marine Littoral Regiment

Core METs:
MCT 1.12.8.1	Conduct Expeditionary Advanced Based Operations (EABO)
MCT 2.10	Support Maritime Domain Awareness (MDA)
MCT 3.2.7.5	Attack Enemy Maritime Targets
MCT 3.2.8	Conduct Expeditionary Strike
MCT 5.14.11	Support Operations in the Information Environment (OIE)
MCT 6.1.1.7.2	Coordinate Air and Missile Defense Actions

Core Plus METs:
MCT 1.6.1	Conduct Offensive Operations
MCT 1.6.4	Conduct Defense Operations
MCT 1.21	Conduct Network Engagement

B.2.2 Regimental Headquarters

Core METs:
MCT 2.10	Support Maritime Domain Awareness (MDA)
MCT 4.11	Plan and Direct Logistics
MCT 5.5.1	Integrate and Operate within a Joint, Interagency, Intergovernmental and Multinational (JIIM) Environment
MCT 5.7.9.1	Command and Control Distributed Maritime Operations
MCT 5.7.9.2	Plan and Direct Littoral Maneuver
MCT 5.7.9.3	Plan and Direct Sea Denial Operations

MCT 5.14.9   Plan and Direct Operations in the Information Environment

Core Plus MET:
MCT 1.15.1.2   Coordinate Foreign Humanitarian Assistance (FHA)

B.2.3 Littoral Combat Team

Core METs:
MCT 1.6.1   Conduct Offensive Operations
MCT 1.6.4   Conduct Defense Operations
MCT 1.12.8.1   Conduct Expeditionary Advanced Based Operations (EABO)
MCT 2.2.15   Conduct Reconnaissance in the Maritime Domain
MCT 3.2.7.5   Attack Enemy Maritime Targets

Core Plus MET:
MCT 1.15   Conduct Civil-Military Operations

B.2.4 Combat Logistics Battalion

Core METs:
MCT 1.12.3   Conduct Prepositioning Operations
MCT 4.1.2   Conduct Ground Supply Operations
MCT 4.2.2   Conduct Ground Equipment Maintenance
MCT 4.3.11   Conduct Littoral Transportation Operations
MCT 4.4   Conduct General Engineering Operations
MCT 4.5   Provide Health Services
MCT 4.13   Conduct Operational Contract Support (OCS)
MCT 6.8   Conduct Explosive Ordnance Disposal Operations

Core Plus METs:
MCT 1.13.2.1   Provide Evacuation Control Center (ECC)
MCT 1.15.1.2   Coordinate Foreign Humanitarian Assistance (FHA)
MCT 1.21   Conduct Network Engagement

B.2.5 Littoral Anti-Air Battalion

Core METs:
MCT 3.2.9   Support Littoral Targeting
MCT 5.3.5.2   Conduct Air Control
MCT 5.3.5.3.2   Conduct Airspace Surveillance
MCT 5.3.5.9   Conduct Air Direction in support of Expeditionary Advanced Base Operations
MCT 6.1.1.8.2   Conduct Short-Range Ground Based Air Defense (GBAD) in support of Maneuver Units

Core Plus METs:
MCT 6.1.1.8.3   Conduct Short-Range GBAD in support of Expeditionary Bases
MCT 6.1.1.8.1   Conduct Medium-Range GBAD in support of Expeditionary Bases

B.3 PROPOSED MARITIME EXPEDITIONARY UNIT MISSION ESSENTIAL TASK LIST

Proposed METL:
MCT 1.6.4.X   Conduct Defense of Naval Task Force

MCT 1.12.1.2	Conduct Amphibious Raid
MCT 1.12.1.8	Conduct Maritime Interception Operations (MIO)
MCT 1.12.1.X	Seize Key Maritime Terrain
MCT 1.12.8.1	Conduct Expeditionary Advanced Based Operations (EABO)
MCT 3.2.8	Conduct Expeditionary Strike
MCT 5.5.1	Integrate/Operate with Joint, Interagency, Intergovernmental and Multinational (JIIM) Organizations
MCT 5.7.9.X	Support Sea Denial Operations
MCT 6.1.7	Conduct Embassy Reinforcement
MCT 6.2.1	Conduct Tactical Recovery of Aircraft and Personnel (TRAP)

B.3.1 MEU Command Element

Proposed METL:

MCT 1.8.3	Conduct Sensitive Site Exploitation
MCT 1.12.1.8	Conduct Maritime Interception Operations (MIO)
MCT 2.1	Plan and Direct Intelligence Operations
MCT 3.2.1.3	Integrate Fire Support with the Scheme of Maneuver
MCT 4.11	Plan and Direct Logistics Operations
MCT 5.5.1	Integrate and Operate with Joint, Interagency, Intergovernmental and Multinational (JIIM) Organizations
MCT 5.7	Exercise Command and Control of All-Domain Forces
MCT 5.7.1	Plan and Direct Amphibious Operations
MCT 5.7.X	Plan and Direct Maritime Operations
MCT 5.7.9.3	Plan and Direct Sea Denial Operations
MCT 5.14.9	Plan and Direct Operations in the Information Environment

B.3.2 MEU Aviation Combat Element

Proposed METL:

MCT 1.3.3.3	Conduct Aviation Operations from Expeditionary Sites
MCT 1.3.4	Conduct Assault Support Operations
MCT 1.12.5.X	Provide Forward Arming and Refueling Points
MCT 2.2.5.2	Conduct Aviation Reconnaissance and Surveillance
MCT 3.2.3	Conduct Aviation Delivered Fires
MCT 5.3.5	Control Aircraft and Missiles
MCT 5.4.1.2	Conduct Electromagnetic Warfare (EW)
MCT 6.1.1.7	Conduct Antiair Warfare (AAW) (Air Defense)
MCT 6.1.1.8	Conduct Active Air and Missile Defense

B.3.3 MEU Ground Combat Element

Proposed METL:

MCT 1.6.1	Conduct Offensive Operations
MCT 1.6.4	Conduct Defensive Operations
MCT 1.12.1.X	Seize Key Maritime Terrain
MCT 2.2.X	Conduct Reconnaissance of Key Maritime Terrain
MCT 3.2.7.5	Attack Enemy Maritime Targets

B.3.4 MEU Logistic Combat Element

Proposed METL:

MCT 1.12.3	Conduct Prepositioning Operations
MCT 4.1.2	Conduct Ground Supply Operations
MCT 4.2.2	Conduct Ground Equipment Maintenance
MCT 4.3.X	Conduct Littoral Transportation Operations
MCT 4.4	Conduct General Engineering Operations
MCT 4.5	Provide Health Services
MCT 4.13	Conduct Operational Contract Support (OCS)
MCT 6.8	Conduct Explosive Ordnance Disposal Ops

APPENDIX C

EXPERIMENT OBJECTIVES

C.1 GENERAL

Experiment objectives provide a framework for the Marine Corps to assess the ability of units to execute operations according to the concepts laid out in *Tentative Manual for Expeditionary Advanced Base Operations*, as well as the validity of the concepts themselves. Collectively, they provide a methodology for assessing the missions and tasks that may be assigned to littoral forces. In addition to missions and tasks associated with the normal warfighting functions, there are specific missions and tasks that are unique to conducting EABO. This appendix aligns experiment objectives and sub-objectives with the EABO missions and tasks to evaluate task accomplishment and mission success.

The experiment objectives are intentionally conceptual, requiring analysis and planning to inform experimentation in EABO. Furthermore, identifying the methods by which tasks and missions are accomplished through experimentation informs future force development and refinement of this manual.

C.2 MISSIONS THAT MAY BE ASSIGNED DURING EABO

- Support sea control operations
- Conduct sea denial operations within the littorals
- Contribute to maritime domain awareness
- Provide forward C5ISRT and counter-C5ISRT
- Provide forward sustainment
- Security Cooperation

C.3 EXPERIMENT OBJECTIVES FOR ASSESSING TASKS AND MISSIONS

Controlled content omitted. The remainder of this appendix can be found at the following web site: https://intelshare.intelink.sgov.gov/sites/mcwl/TMEABOAssessment (SIPRNET).

INTENTIONALLY BLANK

Appendix D
Abbreviations

AADC	area air defense commander
AAW	antiair warfare
ABI	activity-based intelligence
AC2S	aviation command and control system
ACA	airspace control authority
ACE	aviation combat element
ACSA	acquisitioning and cross-servicing agreements
ACSR	aircraft salvage and recovery operations
ADR	airfield damage repair operations
AGS	aviation ground support
AMD	air and missile defense
AMDC	air and missile defense commander
AO	area of operations
AOR	area of responsibility
AREC	air resource element coordinator
ARG	amphibious ready group
ASCOPE	areas, structures, capabilities, organizations, people, and events
ASE	air support element
ASW	antisubmarine warfare
ASWC	antisubmarine warfare commander
ATC	air traffic control
ATO	air tasking order
AVLOG	Marine aviation logistics
BLT	battalion landing team
BRAAT	base recovery after attack operations
C2	command and control
C2D2E	command and control denied or degraded environment
C5ISRT	command, control, communications, computers, combat systems, intelligence, surveillance, reconnaissance, targeting
CA	civil affairs
CAOC	combined air operations center
CCDR	combatant commander
CCF	contingency contracting force
CIEA	classification, identification, and engagement area
CLB	combat logistics battalion
CME	cyberspace mission element
COA	course of action
COLS	concept of logistics support
COMSTRAT	communications strategy and operations
CONEMP	concept of employment
COP	common operating picture
CSG	carrier strike group

CSS	combat service support
CWC	composite warfare commander
DASC	direct air support center
DCA	defensive counterair
DCO	defensive cyberspace operations
DCO-IDM	defensive cyberspace operations-internal defensive measures
DIME	diplomatic, informational, military, economic
DLA	Defense Logistics Agency
DOD	Department of Defense
DODIN	Department of Defense information network
DOS	Department of State
EAB	expeditionary advanced base
EABO	expeditionary advanced base operations
EAF	expeditionary airfield
ECC	Evacuation Control Center
EFR	expeditionary fire and rescue
EMCON	emission control
EMI	electromagnetic interference
EMOE	electromagnetic operation environment
EMS	electromagnetic spectrum
EMSO	electromagnetic spectrum operations
EOD	explosive ordnance disposal
ESG	expeditionary strike group
EW	electromagnetic warfare
EW/C	early warning/control
EXWC	expeditionary warfare commander
F2T2EA	find, fix, track, target, engage, and assess
FAC (A)	forward air controller (airborne)
FARP	forward arming and refueling point
FHA	foreign humanitarian assistance
FMF	Fleet Marine Force
FMFLC	Fleet Marine Force Logistics Command
FSCC	fire support coordination center
FSCM	fire support coordination measure
FSR	field service representative
GBAD	ground based air defense
GCC	Geographic Combatant Commander
GCE	ground combat element
GCPC	Government-wide commercial purchase card
GPN	global positioning network
HEC	helicopter element coordinator
HHQ	higher headquarters
HN	host nation
HNS	host nation support
HSS	health service support
HUMINT	human intelligence

IC	intelligence community
IE	information environment
IPB	intelligence preparation of the battlespace
ISR	intelligence, surveillance, and reconnaissance
IW	irregular warfare
IWC	information operations warfare commander
JAOC	joint air operations center
JFACC	joint force air component commander
JFC	joint force commander
JFLCC	joint force land component commander
JFMCC	joint force maritime component commander
JFSOCC	joint force special operations component commander
JIIM	joint, interagency, intergovernmental, and multinational
JIPOE	joint intelligence preparation of the operational environment
JLEnt	joint logistics enterprise
JP	joint publication
JRFL	joint restricted frequencies list
JTF	joint task force
KLE	key leader engagement
LAAB	littoral antiair battalion
LAAD	low-altitude air defense
LCT	littoral combat team
LFC	littoral force commander
LHA	amphibious assault ship (general purpose)
LHD	amphibious assault ship (multi-purpose)
LMS	littoral maneuver squadron
LOA	littoral operations area
LPD	amphibious transport dock
LRUSV	long-range unmanned surface vessel
LSM	medium landing ship
LTP	littoral transition point
MACC	Marine Corps air command and control agency
MADIS	Marine air defense integrated system
MAGTF	Marine air-ground task force
MAOC	Multi-function Air Operations Center
MARFORPAC	Marine Forces Pacific
MARFORSOC	Marine Forces Special Operations Command
MARLE	Marine liaison element
MASINT	measurement and signature intelligence
MATC CO	Marine air traffic control company
MAW	Marine aircraft wing
MCDP	Marine Corps doctrinal publication
MCISRE	Marine Corps Intelligence, Surveillance, and Reconnaissance Enterprise
MCT	Marine Corps task
MCWP	Marine Corps warfighting publication
MDA	maritime domain awareness

MEF	Marine expeditionary force
MET	mission-essential task
METL	mission-essential task list
MEU	Marine expeditionary unit
MILDEC	military deception
MIO	maritime interception operations
MISO	military information support operations
MLR	Marine littoral regiment
MOC	maritime operations center
MPF	Maritime Prepositioning Force
MWSS	Marine wing support squadron
NALE	naval and amphibious liaison element
NATOPS	Naval Air Training and Operating Procedures and Standardization
NGLS	next-generation logistics ship
NGO	nongovernmental organization
NLRP	nonlethal reference point
NVCHAR	nonlethal vulnerability characteristics
NWP	Navy warfare publication
OA	objective area; operational area
OAS	offensive air support
OCA	offensive counterair
OCO	offensive cyberspace operations
OCS	operational contract support
OE	operational environment
OFSA	own-force signature assessment
OPE	operational preparation of the environment
OPSEC	operational security
OPTASK	operation task; operational tasking (message)
OSINT	open-source intelligence
OST	operational support team
OTC	officer in tactical command
PMESII	political, military, economic, social, infrastructure, and information
PNT	positioning, navigation, and timing
PRP	personnel retrieval and processing
PSYOP	psychological operations (forces)
RADC	regional air defense commander
RFS	request for service; request for sourcing; request for support
RSO&I	reception, staging, onward movement, and integration
SA	surveillance area
SADC	sector air defense commander
SAG	surface action group
SATCOM	satellite communications
SCC	sea combat commander
SIF	stand-in forces
SIGINT	signals intelligence
SIGMAN	signature management

SOF	special operations forces
SOFA	status of forces agreement
SOM	scheme of maneuver
STWC	strike warfare commander
SUPSIT	support situation
SUW	surface warfare
SUWC	surface warfare commander
TAC (A)	tactical air coordinator (airborne)
TACC	tactical air command center (Marine); tactical air control center (Navy)
TACE(A)	tactical air control element airborne
TACON	tactical control
TADC	tactical air direction center
TAOC	tactical air operations center
TE	task element
TF	task force
TG	task group
TIMP	theater infrastructure master plan
TRAP	tactical recovery of aircraft and personnel
TSA	target systems analysis
TSC	theater security cooperation
TTP	tactics, techniques, and procedures
TU	task unit
UAS	unmanned aircraft systems
USG	United States Government
USTRANSCOM	United States Transportation Command
VA	vital area
VI	visual information
WEZ	weapons engagement zone
WRM	war reserve materiel

INTENTIONALLY BLANK

APPENDIX E

GLOSSARY

advanced base — A base located in or near an operational area whose primary mission is to support military operations. (NTRP 1-02 & USMC Dictionary)

advanced naval base — A temporary base established in or near an operational area whose primary purpose is to support fleet operations during the conduct of a naval campaign. (Working definition)

air and missile defense (AMD) — Direct (active and passive) defensive actions taken to destroy, nullify, or reduce the effectiveness of hostile air and ballistic missile threats against friendly forces and assets. (JP 3-01)

air and missile defense commander (AMDC) — Under the composite warfare commander construct, the officer assigned some or all of the officer in tactical command's detailed responsibilities for defensive counterair and granted the tactical control authority to accomplish the assigned missions and tasks. (NTRP 1-02)

airspace control authority (ACA) — The commander designated to assume overall responsibility for the operation of the airspace control system in the airspace control area. (JP 3-52)

alternate position — A position to be occupied when the primary position becomes untenable or unsuitable to carrying out the task. Its position allows fulfillment of the original task. The alternate position is so located that the original task can be accomplished. (USMC Dictionary)

amphibious ready group (ARG) — A Navy task organization formed to conduct amphibious operations, commanded by an amphibious squadron commander. (JP 3-02)

antiair warfare (AAW) — That action required to destroy or reduce to an acceptable level the enemy air and missile threat. Antiair warfare integrates all offensive and defensive actions against enemy aircraft, surface-to-air weapons, and theater missiles into a singular, indivisible set of operations. Antiair warfare is one of the six functions of Marine aviation. (USMC Dictionary)

antisubmarine warfare (ASW) — That segment of naval warfare that involves sensors, weapons, platforms, and targets in the subsurface environment. (NTRP 1-02)

antisubmarine warfare commander (ASWC) — Under the composite warfare commander construct, the officer assigned some or all of the officer in tactical command's detailed responsibilities for antisubmarine warfare and granted the tactical control authority to accomplish the assigned missions and tasks. (NTRP 1-02)

area of operations (AO) — An operational area defined by a commander for land and maritime forces that should be large enough to accomplish their missions and protect their forces. (JP 3-0)

base — A locality from which operations are projected or supported. (JP 1-02)

campaign — A series of related operations aimed at achieving strategic and operational objectives within a given time and space. (JP 5-0)

classification, identification, and engagement area (CIEA) — In maritime operations, the area within the surveillance area and surrounding the vital area(s) in which all objects detected must be classified, identified, and monitored; and the capability maintained to escort, cover, or engage. (NTRP 1-02, USMC Dictionary)

combat logistics battalion (CLB) — A battalion-sized element within a Marine littoral regiment that provides direct support and general support logistics capabilities. (Working definition)

combat logistics force (CLF) — A force that includes both active Navy ships and those operated by the Military Sealift Command within the Naval Fleet Auxiliary Force that carry a broad range of stores, including fuel, food, repair parts, ammunition, and other essential materiel to keep naval forces operating at sea for extended periods. (NTRP 1-02)

composite warfare commander (CWC) — An officer to whom the officer in tactical command of a naval task organization may delegate authority to conduct some or all of the offensive and defensive functions of the force. (JP 3-32)

concept of logistics support (COLS) — A verbal or graphic statement, in a broad outline, of how a commander intends to support and integrate with a concept of operations in an operation or campaign. (JP 4-0)

contested environment — An operational environment that encompasses both the uncertain and hostile environments as defined in joint doctrine. (Working definition derived from JP 3-0)

cooperative security location (CSL) — A facility located outside the United States and US territories with little or no permanent US presence, maintained with periodic service, contractor, or host nation support. Cooperative security locations provide contingency access, logistic support, and rotational use by operating forces and are a focal point for security cooperation activities. (CJCS CM-0007-05)

cover — 1. A type of security operation that protects the force from surprise, develops the situation, and gives commanders time and space in which to respond to the enemy's actions. 2. A form of security operation whose primary task is to protect the main body by fighting to gain time while also observing and reporting information and preventing enemy ground observation of and direct fire against the main body. 3. Offensive or defensive actions to protect the force. (USMC Dictionary)

defensive cyberspace operations (DCO) — Missions to preserve the ability to utilize blue cyberspace capabilities and protect data, networks, cyberspace-enabled devices, and other designated systems by defeating on-going or imminent malicious cyberspace activity. (JP 3-12)

distribution — 1. The arrangement of troops for any purpose, such as a battle, march, or maneuver. 2. The operational process of synchronizing all elements of the logistics system to deliver the right things, to the right place, at the right time, to support the geographic combatant commander. (JP 4-0)

electromagnetic warfare (EW) — Military action involving the use of electromagnetic and directed energy to control the electromagnetic spectrum or to attack the enemy. (JP 3-13.1)

engagement area — An area where the commander intends to contain and destroy an enemy force with the effects of massed weapons and supporting systems. (USMC Dictionary)

expedition — A military operation conducted by an armed force to accomplish a specific objective in a foreign county. (MCDP 3)

expeditionary advanced base (EAB) — An austere, temporary location within a potential adversary's WEZ that provides sufficient maneuver room to accomplish assigned missions seaward while also enabling sustainment and defense of friendly forces therein. (Working definition)

expeditionary advanced base operations (EABO) — A form of expeditionary warfare that involves the employment of mobile, low-signature, persistent, and relatively easy to maintain and sustain naval expeditionary forces from a series of austere, temporary locations ashore or inshore within a contested or potentially contested maritime area in order to conduct sea denial, support sea control, or enable fleet sustainment. (Working definition)

expeditionary force — An armed force organized to accomplish a specific objective in a foreign country. (JP 3-0)

expeditionary strike group (ESG) — An amphibious ready group/Marine expeditionary unit, supported by other forces and led by an embarked Navy flag officer or Marine Corps general officer and an associated command element staff. An expeditionary strike group provides a greater range of amphibious and/or expeditionary warfare planning capabilities for the execution of a variety of missions in the operational environment, including the ability to conduct and support operations ashore and function as a sea base. (NTRP 1-02)

expeditionary warfare — The projection of naval forces into, and their employment within or from, a foreign country and adjacent waters to accomplish a specific mission. (NDP-1)

expeditionary warfare commander (EXWC) — Under the composite warfare commander construct, the officer assigned some or all of the officer in tactical command's detailed responsibilities for expeditionary warfare and granted the tactical control authority to accomplish the assigned missions and tasks. (Working definition derived from NWP 3-56.)

explosive ordnance disposal (EOD) — The process to detect, locate, access, diagnose, render safe / neutralize, recover, exploit, and dispose of explosive or improvised explosive threats. (JP 3-42)

fire support coordination measure (FSCM) — A measure employed by commanders to facilitate the rapid engagement of targets and simultaneously provide safeguards for friendly forces. (JP 3-0)

fleet — An organization of ships, aircraft, Marine Corps forces, and shore-based fleet activities under a commander who may exercise operational, as well as administrative, control. (JP 3-32)

global positioning network (GPN) – An integrated afloat/ashore network of materiel enabling day-to-day campaigning, rapid response to crisis and contingency, and deterrence. (Working definition)

guard — 1. To protect the main force by fighting to gain time while also observing and reporting information. 2. A form of security operation whose primary task is to protect the main force by fighting to gain time while also observing and reporting information, and to prevent enemy ground observation of and direct fire against the main body by reconnoitering, attacking, defending, and delaying. A guard force normally operates within the range of the main body's indirect fire weapons. (USMC Dictionary)

hide — The positioning of a vehicle, individual, or unit so that no part is exposed to observation or direct fire. (USMC Dictionary)

hostile environment — An operational environment in which hostile forces have control, intent, and capacity to effectively oppose or react to the operations a unit intends to conduct. (JP 3-0)

human intelligence (HUMINT) — A category of intelligence derived from information collected and provided by human sources. (JP 2-0)

information environment (IE) — 1. The aggregate of individuals, organizations, and systems that collect, process, disseminate, or act on information. (JP 3-13) 2. The global competitive space that spans the warfighting domains, where all operations depend on information. (MCDP 8)

information operations (IO) — The integrated employment, during military operations, of information-related capabilities in concert with other lines of operation to influence, disrupt, corrupt, or usurp the decision making of adversaries and potential adversaries while protecting our own. (JP 3-13)

information operations warfare commander (IWC) — The officer responsible to the OTC for creating effects and operationally desirable conditions in order to influence, disrupt, corrupt, or usurp the decision making of adversaries and potential adversaries while protecting friendly forces and to assess the information environment to support warfare commanders' objectives in accordance with OTC/CWC direction. (Working definition derived from NWP 3-65)

information warfare (IW) — The integrated employment of Navy's information-based capabilities to degrade, deny, deceive, or destroy an enemy's warfighting capabilities, or to enhance the effectiveness of friendly operations across all domains. (NTRP 1-02)

information-related capability (IRC) — A tool, technique, or activity employed within a dimension of the information environment that can be used to create effects and operationally desirable conditions. (JP 3-13)

intelligence preparation of the battlespace (IPB) — The analytical methodologies employed by the Services or joint force component commands to reduce uncertainties concerning the enemy, environment, time, and terrain. See also **joint intelligence preparation of the operational environment**. (JP 2-01.3)

intermediate force capabilities - Devices used below lethal force options that temporarily impair, disrupt, delay, or prevent threatening actions. (Draft definition)

irregular warfare (IW) - A violent struggle among state and non-state actors for legitimacy and influence over the relevant population(s). (JP 1)

joint intelligence preparation of the operational environment (JIPOE) — The analytical process used by joint intelligence organizations to produce intelligence estimates and other intelligence products in support of the joint force commander's decision-making process. (JP 2-01.3)

key maritime terrain — Any landward portion of the littoral that affords a force controlling it the ability to significantly influence events seaward. (EABO Concept)

littoral — The littoral comprises two segments of operational environment: 1. Seaward: the area from the open ocean to the shore, which must be controlled to support operations ashore. 2. Landward: the area inland from the shore that can be supported and defended directly from the sea. (JP 2-01.3)

littoral antiair battalion (LAAB) — A battalion-sized element within a Marine littoral regiment that provides antiair, forward arming and refueling, and air control capabilities. (Working definition)

littoral combat force (LCF) — A task-organized Navy-Marine Corps formation that is composed of two or more littoral combat groups. (Working definition)

littoral combat group (LCG) — A task-organized Navy-Marine Corps formation that may combine an amphibious ready group with embarked Marine expeditionary unit with a surface action group and other capabilities in order to accomplish specific missions. (Working definition)

littoral combat team (LCT) — A battalion-sized element within a Marine littoral regiment that provides ground combat and surface warfare capabilities. (Working definition)

littoral force (LF) — For the purpose of this tentative manual, integrated and task-organized Marine and Navy units that project naval power within and from expeditionary advanced bases (EABs) by fusing their landward and seaward roles.

littoral force commander (LFC) — A conceptual term, versus a formal title, for the officer who commands all forces within a littoral operations area. (Working definition)

littoral operations area (LOA) — A geographical area of sufficient size for conducting necessary sea, air, and land operations in order to accomplish assigned mission(s) therein. (Working definition)

littoral transition point (LTP) — A designated location where forces conducting surface littoral maneuver will shift from waterborne to overland movement or from overland to back to waterborne movement. (Working definition)

long-range unmanned surface vessel (LRUSV) — A remote-controlled, rigid-hulled, inflatable boat that can carry and launch expendable unmanned aerial systems. (Working definition)

Marine air-ground task force (MAGTF) — The Marine Corps' principal organization for all missions across the range of military operations, composed of forces task organized under a single commander capable of responding rapidly to a contingency anywhere in the world. The types of forces in the MAGTF are functionally grouped into four core elements: a command element, an aviation combat element, a ground combat element, and a logistics combat element. The four core elements are categories of forces, not formal commands. The basic structure of the MAGTF never varies, though the number, size, and type of Marine Corps units composing each of its four elements will always be mission dependent. The flexibility of the organizational structure allows for one or more subordinate MAGTFs to be assigned. In a joint or multinational environment, other service or multinational forces may be assigned or attached. (USMC Dictionary)

Marine littoral regiment (MLR) — A Marine Corps formation designed to persist within an adversary's weapons-engagement zone in order to conduct expeditionary advanced base operations in support of fleet operations. (Working definition)

maritime domain — The oceans, seas, bays, estuaries, islands, coastal areas, and the airspace above these, including the littorals. (JP 3-32)

maritime domain awareness (MDA) — The effective understanding of anything associated with the maritime domain that could impact the security, safety, economy, or environment of a nation. (JP 3-32)

maritime interception operations (MIO) — Efforts to monitor, query, and board merchant vessels in international waters to enforce sanctions against other nations, such as those in support of United Nations Security Council resolutions, and/or prevent the transport of restricted goods. (JP 3-03)

Medium Landing Ship (LSM) - The primary mission of LSM is to provide intra-theater maneuver and mobility of Naval Expeditionary Forces from shore-to-shore in uncertain and contested environments. Beachable but not a forcible entry platform. Not a replacement for amphibious warfare ships or landing craft. (Working definition)

military information power — The total means of force or information capability applied against a relevant actor to enhance lethality, survivability, mobility, or influence. (DC I and DC CD&I joint memorandum dtd 22 Jan 2020).

military information support operations (MISO) — Planned operations to convey selected information and indicators to foreign audiences to influence their emotions, motives, objective reasoning, and ultimately the behavior of foreign governments, organizations, groups, and individuals in a manner favorable to the originator's objectives. (JP 3-13.2)

mine countermeasures (MCM) — All methods for preventing or reducing damage or danger from mines. (JP 3-15)

naval integration — Synchronization of the complementary authorities, capabilities, capacities, roles, investments, processes, systems, and authorities of Navy, Marine Corps, and Coast Guard headquarters and subordinate units to expand the ability of task-organized naval forces to deliver effects in all domains and across the competition continuum. (Working definition)

network engagement — Interactions with friendly, neutral, and threat networks, conducted continuously and simultaneously at the tactical, operational, and strategic levels, to help achieve the commander's objectives within an operational area. (JP 3-25)

noncombatant evacuation operation (NEO) — An operation whereby noncombatant evacuees are evacuated from a threatened area abroad, which includes areas facing actual or potential danger from natural or manmade disaster, civil unrest, imminent or actual terrorist activities, hostilities, and similar circumstances, that is carried out with the assistance of the Department of Defense. (JP 3-68)

nonlethal weapons (NLW) - A weapon, device, or munition that is explicitly designed and primarily employed to incapacitate personnel or materiel immediately, while minimizing fatalities, permanent injury to personnel, and undesired damage to property in the target area or environment. (JP 3-09)

offensive cyberspace operations (OCO) — Missions intended to project power in and through cyberspace. (JP 3-12)

officer in tactical command (OTC) — In maritime usage, the senior officer present eligible to assume command, or the officer to whom the senior officer has delegated tactical command. (JP 3-32)

operational area (OA) — An overarching term encompassing more descriptive terms (such as area of responsibility and joint operations area) for geographic areas in which military operations are conducted. (JP 3-0)

operational control (OPCON) — The authority to perform those functions of command over subordinate forces involving organizing and employing commands and forces, assigning tasks, designating objectives, and giving authoritative direction necessary to accomplish the mission. (JP 1)

operational environment (OE) — A composite of the conditions, circumstances, and influences that affect the employment of capabilities and bear on the decisions of the commander. (JP 3-0)

operational preparation of the environment (OPE) — The conduct of activities in likely or potential areas of operations to prepare and shape the operational environment. (JP 3-05)

operational tasking (message) (OPTASK) — Maritime-unique formatted message used by both the United States Navy and NATO to provide detailed information for specific aspects within individual areas of warfare and for tasking resources. This includes logistics, may be issued at all levels above the unit, and may be Navy-wide or focused on a particular theater or strike group. (NTRP 1-02)

position — 1. A location or area occupied by a military unit. 2. The location of a weapon, unit, or individual from which fire is delivered upon a target. (USMC Dictionary)

prevention of mutual interference (PMI) — In submarine operations, procedures established to prevent submerged collisions between friendly submarines, between submarines and friendly surface ship towed bodies and arrays, and between submarines and any other hazards to submerged navigation. (JP 3-32)

primary position — A position that provides the best means to accomplish the assigned mission. (USMC Dictionary)

screen — 1. A security element whose primary task is to observe, identify, and report information, and only fight in self-protection. 2. A form of security operation that primarily provides early warning to the protected force. (USMC Dictionary)

sea base — An inherently maneuverable, scalable aggregation of distributed, networked platforms that enables the global power projection of offensive and defensive forces from the sea and includes the ability to assemble, equip, project, support, and sustain those forces without reliance on land bases within the joint operations area. (NTRP 1-02)

sea basing — The deployment, assembly, command, projection, reconstitution, sustainment, and re-employment of joint power from the sea without reliance on land bases within the operational area. (JP 3-02)

sea combat commander (SCC) — Under the composite warfare commander construct, the officer assigned some or all of the officer in tactical command's detailed responsibilities for sea combat and granted the tactical control authority to accomplish the assigned missions and tasks; this is an optional position within the composite warfare commander structure. (NTRP 1-02)

sea control — The condition in which one has freedom of action to use the sea for one's own purposes in specified areas and for specified periods of time and, where necessary, to deny or limit its use to the enemy. Sea control includes the airspace above the surface and the water volume and seabed below. (NTRP 1-02)

sea denial — The ability to partially or completely denying the adversary the use of the sea with a force that may be insufficient to ensure the use of the sea by one's own forces. (NTRP 1-02)

sector — An area designated by boundaries within which a unit operates and for which it is responsible. (NTRP 1-02, MCRP 5-2A)

signature control (SIGCON) — The management and assessment of observable and measurable force signatures and profiles across all domains and spectrums to deny, degrade, or deceive the adversary's ability to detect and engage the strike force. SIGCON affects the ability of an adversary's sensor(s) to detect friendly forces by understanding and controlling activities within exploitable domains. SIGCON is a discipline focused on affecting the adversary's "left of kill chain" ISRT capabilities. (TM 3-13.1-17)

signature management (SIGMAN) — The process by which we understand own-force signatures and indicators; identify adversary methods and capabilities to collect and analyze those signatures; develop and implement countermeasures to mask those signatures; develop and implement, when necessary, methods to project false signatures that protect friendly forces from adversary exploitation or draw the adversary toward a specific course of action or position of disadvantage. (*Marine Corps Concept for Signature Management*)

stand-off engagement capabilities — Long-range capabilities designed to be employed from outside the effective range of an opponent's weapons in order to minimize risk to one's own forces. (Working definition)

stand-in forces (SIF) — Those mobile, low-signature, persistent, and relatively easy to maintain and sustain naval expeditionary forces designed to persist and operate inside a competitor's weapons-engagement zone to cooperate with partners, support host nation sovereignty, confront malign behavior, and, in the event of conflict, engage the enemy in close-range battle. (SIF concept)

strike — An attack to damage or destroy an objective or a capability. (JP 3-0)

strike warfare (STW) — Naval operations to destroy or neutralize enemy targets ashore, including attacks against strategic targets, such as manufacturing facilities and operating bases, from which the enemy is capable of conducting or supporting air, surface, or subsurface operations against friendly forces. (NTRP 1-02)

strike warfare commander (STWC) — Under the composite warfare commander construct, the officer assigned some or all of the officer in tactical command's detailed responsibilities for strike warfare and granted the tactical control authority to accomplish the assigned missions and tasks. (NTRP 1-02)

support — 1. The action of a force that aids, protects, complements, or sustains another force in accordance with a directive requiring such action. 2. A unit that helps another unit in battle. 3. An element of a command that assists, protects, or supplies other forces in combat. (JP 1)

support situations (SUPSITs) — The degree, manner, and duration of the action of a maritime task organization or portion thereof, which aids, protects, complements, or sustains any other maritime task organization when a support command relationship is not established. (NWP 3-56)

SUPSIT Alpha — The supporting force is to join and integrate with the supported force. The senior officer present, or the officer to whom he/she has delegated tactical command, becomes the OTC of the integrated force. (NWP 3-56)

SUPSIT Bravo — The supporting force does not integrate with the supported force. Unless otherwise ordered, the supported commander of the two forces is to coordinate the tactical operations of the two forces. (NWP 3-56)

SUPSIT Charlie — The supporting force commander has discretion on how best to provide support. (NWP 3-56)

surface action group (SAG) — A temporary or standing organization of combatant ships, other than carriers, tailored for a specific tactical mission. (JP 3-32)

surface warfare (SUW) — That portion of maritime warfare in which operations are conducted to destroy or neutralize enemy naval surface forces and merchant vessels. (JP 3-32)

surface warfare commander (SUWC) — Under the composite warfare commander construct, the officer assigned some or all of the officer in tactical command's detailed responsibilities for surface warfare and granted the tactical control authority to accomplish the assigned missions and tasks. (NTRP 1-02)

surveillance area (SA) — In surface warfare, the area in the operational environment that extends out to a range that equals the force's ability to conduct a systematic observation of a surface area using all available and practical means to detect any vessel of possible military concern. (NTRP 1-02)

tactical control (TACON) — The authority over forces that is limited to the detailed direction and control of movements or maneuvers within the operational area necessary to accomplish missions or tasks assigned. (JP 1)

tactical recovery of aircraft and personnel (TRAP) — A Marine Corps mission performed by an assigned and briefed aircrew for the specific purpose of the recovery of personnel, equipment, and/or aircraft when the tactical situation precludes search and rescue assets from responding and when survivors and their location have been confirmed. (JP 3-50)

vital area (VA) — A designated area or installation to be defended by air defense units. (NTRP 1-02)

waterspace management (WSM) — The allocation of waterspace in terms of antisubmarine warfare attack procedures to permit the rapid and effective engagement of hostile submarines while preventing inadvertent attacks on friendly submarines. (JP 3-32)

weapons engagement zone (WEZ) — 1. In air and missile defense, airspace of defined dimensions within which the responsibility for engagement of air threats normally rests with a particular weapon system. (JP 3-01) 2. In antisubmarine warfare, the area defined by a submarine datum expanded by a predicted furthest-on-circle and the maximum effective torpedo firing range (for a torpedo threat) or 3.

The maximum effective missile firing range (for an antiship cruise missile threat). (NTRP 1-02) 4. The maximum range at which a combatant can detect adversary forces and effectively employ anti-ship missiles and land-attack missiles against them. (Working definition)

INTENTIONALLY BLANK

www.ingramcontent.com/pod-product-compliance
Lightning Source LLC
Chambersburg PA
CBHW082041230426
43670CB00016B/2733